# Energy and Life

CAMBRIDGE UNIVERSITY PRESS

*Cambridge*

*London   New York   New Rochelle*

*Melbourne   Sydney*

# Contents

# Preface

The Inner London Education Authority's Advanced Biology Alternative Learning (ABAL) project has been developed as a response to changes which have taken place in the organisation of secondary education and the curriculum. The project is the work of a group of biology teachers seconded from ILEA secondary schools. ABAL began in 1978 and since then has undergone extensive trials in schools and colleges of further education. The materials have been produced to help teachers meet the needs of new teaching situations and provide an effective method of learning for students.

Teachers new to A-level teaching or experienced teachers involved in reorganisation of schools due to the changes in population face many problems. These include the sharing of staff and pupils between existing schools and the variety of backgrounds and abilities of pupils starting A-level courses whether at schools, sixth form centres or colleges. Many of the students will be studying a wide range of courses, which in some cases will be a mixture of science, arts and humanities.

The ABAL individualised learning materials offer a guided approach to A-level biology and can be used to form a coherent base in many teaching situations. The materials are organised so that teachers can prepare study programmes suited to their own students. The separation of core and extension work enables the academic needs of all students to be satisfied. Teachers are essential to the success of this course, not only in using their traditional skills, but for organising resources and solving individual problems. They act as personal tutors, and monitor the progress of each student as he or she proceeds through the course.

The materials aim to help the students develop and improve their personal study skills, enabling them to work more effectively and become more actively involved and responsible for their own learning and assessment. This approach allows the students to develop a sound understanding of fundamental biological concepts.

# Acknowledgements

**Figures:** 13, 27, 52, 80, A. Langham; 43, 47, 48, M. Thompson; 44, 56, 81, 83, Biophoto Associates; 49, J.D. Dodge; 53, R.H. Emson.

**Examination questions:** By permission of the Associated Examining Board and the University of London Examining Board.

Cover illustration prepared from a photograph supplied by Ardea Ltd.

# How to use this unit

This is not a textbook. It is a guide that will help you learn as effectively as possible. As you work through it, you will be directed to practical work, audio-visual resources and other materials. There are sections of text in this guide which are to be read as any other book, but much of the guide is concerned with helping you through activities designed to produce effective learning. The following list gives details of the ways in which the unit is organised.

## (1) Objectives

Objectives are stated at the beginning of each section. They are important because they tell you what you should be able to do when you have finished working through the section. They should give you extra help in organising your learning. In particular, you should check after working through each section that you can achieve all the stated objectives and that you have notes which cover them all.

## (2) Self-assessment questions (*SAQ*)

These are designed to help you think about what you are reading. You should always write down answers to self-assessment questions and then check them immediately with those answers given at the back of this unit. If you do not understand a question and answer, make a note of it and discuss it with your tutor at the earliest opportunity.

## (3) Summary assignments

These are designed to help you make notes on the content of a particular section. They will provide a useful collection of revision material. They should therefore be carried out carefully and should be checked by your tutor for accuracy. If you prefer to make notes in your own way, discuss with your tutor whether you should carry out the summary assignments.

## (4) Self tests

There are one or more self tests for each section. They should be attempted a few days after you have completed the relevant work and not immediately after. They will help you identify what you have not understood or remembered from a particular section. You can then remedy any weaknesses identified. If you cannot answer any questions and do not understand the answers given, then check with your tutor.

## (5) Tutor assesssed work

At intervals through the unit, you will meet an instruction to show work to your tutor. This will enable your tutor to monitor your progress through the unit and to see how well you are coping with the material. Your tutor will then know how best to meet your individual needs.

## (6) Past examination questions

At various points in the unit you will come across past examination questions. These are only included where they are relevant to the topic under study and have been selected both to improve your knowledge of that topic and also to give you practice in answering examination questions.

## (7) Audio-visual material

A number of activities in this unit refer to video cassettes which may be available from your tutor. They deal with topics which cannot be covered easily in text or practical work, as well as providing a change from the normal type of learning activities. This should help in motivating you.

## (8) Extension work

This work is provided for several reasons: to provide additional material of general interest, to provide more detailed treatment of some topics, to provide more searching questions that will make demands on your powers of thinking and reasoning.

## (9) Practicals

These are an integral part of the course and have been designed to lead you to a deeper understanding of the factual material. You will need to organise your time with care so that you can carry out the work suggested in a logical sequence. If your A-level examination requires your practical notebook to be assessed, you must be careful to keep a record of this work in a separate book. A hazard symbol, ☠, is used in the Materials and Procedures sections to mark those substances and procedures which must be treated with particular care.

## (10) Discussions

Talking to one another about biological ideas is a helpful activity. To express yourself in your own words, so that others can understand you, forces you to clarify your thoughts. When a sufficient number of your class (at least three, but not more than five) have covered the material indicated by a discussion instruction, you should have a group discussion. Question individuals if what they say is not clear. This is the way that you will both learn and understand.

## (11) Programmed learning texts

There is one programmed learning text in this unit. To use it most effectively, you should take a piece of paper or card (that is a mask) and place it on the first page of the programmed text so that only the text above the first horizontal dividing line is visible. Such an item is called a frame. Each frame will provide you with information and ask you to make a response. When you have done this, move the mask down to reveal the correct answer and the next frame. This procedure should be repeated throughout the frame.

## (12) Post-test

A post-test is available from your tutor when you finish this unit. This will be based on past examination questions and will give you an idea of how well you have coped with the material in this unit. It will also indicate which areas you should consolidate before going on to the next unit.

## Study and practical skills

The ABAL introductory unit *Inquiry and investigation in biology* introduced certain study and practical skills which will be practised and improved in this unit. These included
(a) the QS3R method of note-taking;
(b) the construction of graphs, histograms and tables;
(c) the analysis of data;
(d) drawing of biological specimens;
(e) use of the light microscope;
(f) the design of practical investigations;
(g) comprehension of written reports;
(h) discussion groups.

1   The relationship of themes in this unit

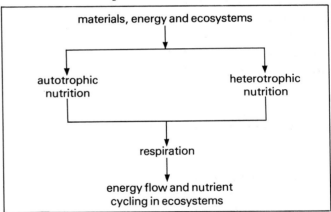

## Pre-knowledge for this unit

You should have an understanding of the following.
The nature of energy
Units of measurement of energy
First and second laws of thermodynamics
Role of ATP and ADP in metabolism
Structure of carbohydrates, lipids and protein and their function in living organisms

These topics are covered in the unit *Cells and the origin of life*.

# Introduction to this unit

All living organisms require a source of energy from the environment in order to stay alive. Energy is necessary for growth and development and for activities such as transport of substances within the body, elimination of waste materials, movement, nerve action and reproduction. In addition to energy, living organisms also require a supply of basic materials.

The method of obtaining both energy and basic materials is closely linked but varies for plants and animals.

Green plants use light energy from the sun and simple inorganic materials from the environment. Animals, some plants, and fungi depend on other organisms for their energy and materials. These two methods of obtaining energy and materials are known respectively as **autotrophic** and **hetero-trophic nutrition**. Both types are found among the microorganisms. You will study autotrophic and heterotrophic nutrition in sections 2 and 3.

The energy acquired by the organism is usually stored in molecules of carbohydrate or fat. During the process of respiration, this energy can be released and used for various activities such as those described above. Section 4 explains the way in which this is brought about.

The internal processes of living organisms do not take place in isolation from the whole organism and its environment. Sections 1 and 5 will help you to see the relationship between energy and life in an ecological context.

Like any other animal, man obtains his energy and basic materials heterotrophically and this energy is released for use during respiration. However, the relationship between man and energy is not just a simple biological one.

The video sequence entitled *Man and energy* introduces this unit by examining the important role energy plays in human societies.

---

## AV 1: Man and energy
---

### Materials

VCR and monitor
ABAL video sequence: *Man and energy*
Worksheets

### Procedure

(*a*) Check that you have all the relevant materials for this activity.

(*b*) Check that the video cassette is set up ready to show the appropriate sequence – *Man and energy*.

(*c*) Start the VCR and stop it to complete the worksheets as instructed.

(*d*) If you do not understand something, stop the video, rewind, and study the relevant material again before consulting with your tutor.

(*e*) If possible, work through the video and worksheets with a small group and discuss the material with your fellow students.

# Section 1 Materials, energy and ecosystems

## 1.1 Introduction and objectives

In this section you will examine nutrition and consider how different types of organisms show different forms of nutrition. You will also see how these different types of organisms interact with each other in their natural environment.

After completing this section, you should be able to do the following.

(a) Define or recognise definitions of the following terms:

| | | |
|---|---|---|
| abiotic | ecosystem | omnivore |
| autotrophic | energy | parasite |
| biotic | food chain | population |
| carnivore | food web | producer |
| community | grazing chain | saprophyte |
| consumer | herbivore | scavenger |
| decay chain | heterotrophic | trophic level |
| decomposer | nutrient | |
| detritivore | nutrition | |

(b) Make lists of
(i) six examples of ecosystems,
(ii) six abiotic components of ecosystems.

## 1.2 Nutrition

**Nutrition** is the sum of the processes by which organisms obtain the basic materials and energy necessary for growth and repair, reproduction, movement and the production of various secretions (enzymes, hormones, etc.).

The basic materials are called **nutrients**, and they can be divided into two general categories:
(a) **inorganic nutrients** such as water and mineral salts obtained directly from the physical environment;
(b) **organic nutrients** such as carbohydrates, proteins and fats; these are manufactured by green plants from inorganic materials. The energy required for the synthesis of organic molecules comes from the sun.

Animals are unable to manufacture organic nutrients from inorganic nutrients and thus are, directly or indirectly, dependent upon plants for their nutrients.

*SAQ 1* Choose one of the following words to fit the definitions below.

nutrition, nutrients, energy

(a) The ability to do work.
(b) The materials required from the environment by living organisms.
(c) The processes whereby living organisms obtain raw materials and energy from the environment.

*SAQ 2* For the six commodities listed below, state whether they may act as an energy supply, inorganic nutrient, or organic nutrient.
(a) sunlight        (d) calcium
(b) fat             (e) carbon dioxide
(c) water           (f) protein

## 1.3 Ecology and ecosystems

The environment of an organism includes not only its physical surroundings but also other living organisms. The branch of biology which studies the relationships between organisms and their environment is **ecology**. Nutrition clearly forms the basis of many of these relationships.

Ecological studies can be made at different levels. At one end of the scale, the ecologist can look at the interactions of a single organism with its environment. At the other extreme, the largest ecological unit is the **biosphere**, which may be defined as that part of the planet which is occupied by living organisms. The biosphere, as a whole, is far too large and complex for detailed analysis, and ecologists generally study one small part of it at a time. Some examples of levels of complexity between those of individual organisms and biosphere are given in figure 2.

### 2 Levels of ecological investigation

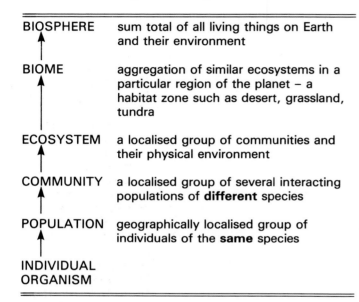

| BIOSPHERE | sum total of all living things on Earth and their environment |
| --- | --- |
| BIOME | aggregation of similar ecosystems in a particular region of the planet – a habitat zone such as desert, grassland, tundra |
| ECOSYSTEM | a localised group of communities and their physical environment |
| COMMUNITY | a localised group of several interacting populations of **different** species |
| POPULATION | geographically localised group of individuals of the **same** species |
| INDIVIDUAL ORGANISM | |

Many ecologists choose an ecosystem as their unit of study. An **ecosystem** is a natural unit of all the plants, animals and microorganisms in an area together with their non-living environment. Ecosystems are very variable in size, ranging from a small puddle to a large forest. They contain non-living or **abiotic** components together with living or **biotic** components. These components interact with each other in a great variety of ways. This section examines the interactions associated with nutrition, starting by looking at a particular ecosystem, a freshwater pond.

*SAQ 3* Which of the following are examples of ecosystems?
(*a*) a hive of bees
(*b*) a lake
(*c*) a meadow
(*d*) a sandy shore
(*e*) a shoal of fish

### AV 2: The pond ecosystem

This section will enable you to learn more about ecosystems by looking in more detail at one particular example.

**Materials**

VCR and monitor
ABAL video sequence: *The pond ecosystem*
Worksheets

**Procedure**

(*a*) Check that you have all the relevant materials for this activity.

(*b*) Check that the video cassette is set up ready to show the appropriate sequence – *The pond ecosystem*.

(*c*) Start the VCR and stop it to complete the worksheets as instructed.

(*d*) If you do not understand anything, stop the video, rewind, and study the relevant material again before consulting with your tutor.

(*e*) If possible, work through the video and worksheets with a small group and discuss the material with your fellow students.

### AV 3: Analysing feeding relationships in a woodland ecosystem

In this section you will study the feeding relationships between organisms in an ecosystem using a woodland ecosystem as an example.

**Materials**

Poster entitled *Oak woods*

**Procedure**

The poster shows some of the organisms typical of an oak-dominated woodland. Study it carefully.

*SAQ 4* Using information from the poster, construct a food web for the oak wood. The starting points of the food chains are given in figure 3.

**3  The starting points of some oak wood food chains**

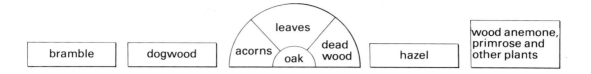

**4  Classification of heterotrophs**

In cases where the food of a particular animal is not given on the poster, try to find this information out for yourself by consulting suitable books such as those listed on page 00.

Indicate the feeding relationships between organisms by means of straight arrows, as in this example:

$$grass \longrightarrow rabbit \longrightarrow fox$$

In spite of the apparently confusing complexity of the food web diagram, there are several ways it can be analysed. Firstly, individual food chains may be considered. Refer to figure 126 and answer the following questions.

*SAQ 5* (a) List three food chains with four components.
(b) List three food chains with three components.
(c) List three animals found at the end of food chains.
(d) Which three animals are found at the end of the greatest number of food chains?

A second way of analysing complex food webs is to classify the various categories of organisms according to their mode of nutrition. The most obvious basis for a classification of this type is the autotroph/heterotroph distinction. Heterotrophs can be sub-divided into six categories according to the nature of their food supply, as shown in figure 4.

*SAQ 6* Study figure 126 and state in which of the categories shown in figure 4 you would place the following organisms.

stag beetle, chiffchaff, fungi

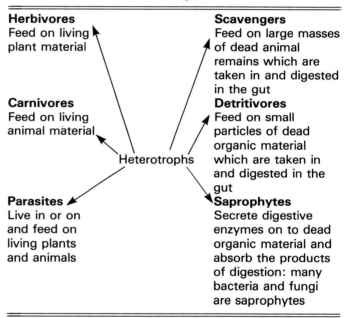

The food chains arising from living plants are often referred to as **grazing chains**. Food chains arising from dead material are known as **decay chains**.

In addition to breaking down the organic molecules contained in the dead bodies of other organisms or their wastes, the digestive activity of saprophytes also produces inorganic materials such as water and mineral salts, which return to the abiotic environment. In everyday language, the results of the activity of saprophytes is known as **putrefaction** or **decay**.

The saprophytes, together with the detritivores, form a vital community of organisms in an ecosystem and, to emphasise their important role, ecologists often refer to them collectively as the **decomposers**.

## 1.4 Trophic levels

All food chains start with autotrophs, generally green plants. The heterotrophs are dependent on the plants for their supplies of these energy-yielding molecules. For this reason, the green plants are collectively referred to as the **producers**.

Animals depend directly or indirectly on the producers for their food. They are known as **consumers**. Herbivores, which feed directly on plants, are called **primary consumers**. Carnivores, which feed on the herbivores, are called **secondary consumers**. Carnivores feeding on other carnivores are called **tertiary consumers**, and so on.

These groupings are referred to as **trophic (feeding) levels**. The producers form the first trophic level, the primary consumers the second, and so on.

Members of each level eventually die and form the food of the decomposers. This is summarised in figure 5.

5   Trophic levels

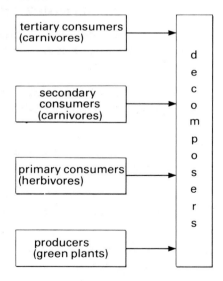

**SAQ 7** Classify the organisms in the oak wood ecosystem (figure 126) into the following categories.
   producers, primary consumers, secondary consumers, tertiary consumers, decomposers

**SAQ 8** What is meant by the term trophic level? Give two examples.

## 1.5 References for section 1

*Ecological Energetics* by John Phillipson, Studies in Biology No. 1.
*Fungal Saprophytism* by Harry J. Hudson, Studies in Biology No. 32.
*Ecology of Fungi* by R. K. Robinson, chapter 2.

## 1.6 Summary assignment for section 1

1 Write definitions and give four examples of the following terms.
   ecosystem, food chain, nutrient

2 Write definitions of the following terms.
   abiotic, biotic, population, community, food web, nutrition, producer, primary consumer, secondary consumer, autotroph, heterotroph, scavenger, detritivore, saprophyte

Show this work to your tutor.

Self test 1, page 97, covers section 1 of this unit.

# Section 2  Autotrophic nutrition

## 2.1 Introduction and objectives

In this section you will study autotrophic nutrition. You will begin by examining the developments which led to an understanding of photosynthesis and those factors which are known to affect the process. Then you will study the process itself with its two main phases – the light-dependent and the light-independent phases. Next, you will consider the whole plant and examine those structures in which photosynthesis occurs. Finally, you will examine the importance of minerals for plants.

At the end of this section you should be able to do the following.

(*a*) Describe the contribution of the following scientists in investigating the raw materials and products of photosynthesis.

> Jean Baptiste van Helmont, Joseph Priestley, Jan Ingen-Housz, Nicholas Theodore de Saussure

(*b*) Write an overall equation for the process of photosynthesis, and describe this process in one sentence.

(*c*) Define or recognise definitions of the following terms: absorption spectra, action spectrum, limiting factor, essential element, trace element.

(*d*) List the internal and external factors which affect the rate of photosynthesis.

(*e*) Describe the effects on the rate of photosynthesis of carbon dioxide concentration, light intensity, light wavelength and temperature. Interpret graphs which illustrate these effects.

(*f*) State what is meant by the term compensation point and explain how this varies for different plants.

(*g*) Describe how some plants are adapted to the quality and intensity of the light available in their normal habitat.

(*h*) Describe the contribution of the following scientists to our understanding of the light-dependent reaction.

> Robert Hill, Daniel Arnon *et al*.

(*i*) Outline the two processes involved in the light-dependent reaction and state the raw materials and products for each process.

(*j*) Describe the role of pigment molecules in photosynthesis.

(*k*) Define electron flow.

(*l*) Explain the meaning of cyclic photophosphorylation and non-cyclic photophosphorylation.

(*m*) Describe the role of water in photosynthesis.

(*n*) Describe how the use of isotopes led to the determination of the source of the oxygen released during photosynthesis.

(*o*) Describe the contribution of Melvin Calvin in investigating the light-independent reactions of photosynthesis.

(*p*) Explain what is meant by two-dimensional paper chromatography and autoradiography.

(*q*) Construct a diagram representing the Calvin cycle showing the chemicals involved and the number of carbon atoms per molecule.

(*r*) State those parts of the Calvin cycle in which ATP and $NADPH_2$ are needed.

(*s*) Describe how plants produce fats and proteins as well as carbohydrates from the Calvin cycle.

(*t*) Describe how autotrophic bacteria differ from other green plants in terms of their photosynthetic pigments and their hydrogen source for photosynthesis.

(*u*) Explain how chemosynthetic bacteria obtain their energy. What are the ecological significances of this?

(*v*) Describe the surface structure of a leaf.

(*w*) List and draw representative cells making up a cross-section of a leaf.

(*x*) Describe the relationship between structure and function in the leaf and in the chloroplast.

(*y*) Outline the technique for water culture.

(*z*) List the essential elements for plants. Outline the role and deficiency characteristics of these elements in the plant.

*Extension*

(*a*) Explain the contribution of Blackman to our knowledge of photosynthesis.

(*b*) Describe what is meant by the terms photosystem I and photosystem II and show how these are related to cyclic and non-cyclic electron flow.

(*c*) Distinguish between $C_3$ and $C_4$ plants.

## 2.2 Photosynthesis – an historical perspective

Plant nutrition has been investigated by scientists for centuries because of the importance of growing plants for food. The essential process in plant nutrition is photosynthesis, yet it was not until the nineteenth century that scientists could give an accurate description of the raw materials and products of photosynthesis. This section outlines the major advances made in the investigation of photosynthesis during the period 1648–1804.

### 2.2.1 Jean Baptiste van Helmont (1577–1644)

One of the earliest investigations into the raw materials and products of photosynthesis was carried out by a Dutch physician, Jean Baptiste van Helmont. He did not know about photosynthesis as such, he was simply investigating the raw materials from which vegetable matter was formed. An excerpt

from one of his reports, published in 1648, is printed below.

*By experiment, that all vegetable matter is totally and materially of water alone* by Jean Baptiste van Helmont. From *Ortus Medicinae* pp. 108–109, Amsterdam, 1648

'I took an earthen vessel, in which I put 200 pounds of earth that had been dried in a furnace, which I moistened with rainwater, and I implanted therein the trunk or stem of a willow tree, weighing five pounds. And at length, five years being finished, the tree sprung from then did weigh 169 pounds and about three ounces. When there was need, I always moistened the earthen vessel with rainwater or distilled water, and the vessel was large and implanted in the earth. Lest the dust that flew about should be co-mingled with the earth, I covered the lip or mouth of the vessel with an iron plate covered with tin and easily passable with many holes. I computed not the weight of the leaves that fell off in the four autumns. At length, I again dried the earth of the vessel, and there was found the same 200 pounds, wanting about two ounces. Therefore 164 pounds of wood, bark and root arose out of water only.'

Van Helmont's conclusion could be summarised as a word equation:

**Equation 1**

$$\text{water} \xrightarrow{\text{vegetation}} \text{vegetable matter}$$

*SAQ 9* What did van Helmont overlook when coming to this conclusion? Suggest why he came to such an incorrect conclusion.

### 2.2.2 Joseph Priestley (1733–1804)

The next recorded step forward in scientific investigations into the nutrition of autotrophs was taken in 1772. Joseph Priestley, a British scientist, was actually investigating the effect of vegetation on different kinds of air. He did not, at this stage, appreciate the significance of his observations for the understanding of plant nutrition. An excerpt from one of his reports follows.

*Observations on different kinds of air* by Joseph Priestley, LL D, FRS. Abridged from *Philosophical Transactions of the Royal Society of London*, **62**, pp. 166–170 (1772).

I flatter myself that I have accidentally hit upon a method of restoring air which has been injured by the burning of candles, and that I have discovered at least one of the restoratives which nature employs for this purpose. It is vegetation. In what manner this process in nature operates, to produce so remarkable an effect, I do not pretend to have discovered; but a number of facts declare in favour of this hypothesis. I shall introduce my account of them, by reciting some of the observations which I made on the growing of plants in confined air, which led to this discovery.

One might have imagined that, since common air is necessary to vegetable, as well as to animal life, both plants and animals had affected it in the same manner, and I own I had that expectation when I first put a sprig of mint into a glass-jar, standing inverted in a vessel of water; but when it had continued growing there for some months, I found that the air would neither extinguish a candle, nor was it at all inconvenient to a mouse, which I put into it.
Finding that candles burn very well in air in which plants had grown a long time, and having had some reason to think that there was something attending vegetation, which restored air that had been injured by respiration, I thought it was possible that the same process might also restore the air that had been injured by the burning of candles.
Accordingly, on the 17th of August, 1771, I put a sprig of mint into a quantity of air, in which a wax candle had burned out, and found that, on the 27th of the same month, another candle burned perfectly well in it. This experiment I repeated, without the least variation in the event, not less than eight or ten times in the remainder of the summer. Several times I divided the quantity of air in which the candle had burned out, into two parts, and putting the plant into one of them, left the other in the same exposure, contained, also, in a glass vessel immersed in water, but without any plant; and never failed to find, that a candle would burn in the former, but not in the latter. I generally found that five or six days were sufficient to restore this air, when the plant was in its vigour.'

Priestley's findings could be summarised in a word equation:

**Equation 2**

$$\text{injured air} \xrightarrow{\text{vegetation}} \text{restored air}$$

*SAQ 10* From the preceding account,
(a) was Priestley's work qualitative or quantitative? (Give a reason for your answer.)
(b) describe how this method showed a concern for controls and the repeatability of results.
(c) what was Priestley's hypothesis?

Priestley's discovery was greeted with enthusiasm by scientists of his time because it seemed to provide the answer to a great puzzle. How was it that the bad air being constantly produced by fires and animals did not collect in the atmosphere and eventually dominate it? The answer appeared to be that vegetation cleansed and purified the atmosphere.

For many years after Priestley's discovery numerous investigators tried to repeat his experiments, but were unable to obtain the same results consistently.

The process of validating a scientist's work by other independent scientists is a very important contribution to scientific progress. Before the results of investigations, or theories based on such results, are acceptable to the scientific community, they must be repeated and verified by other scientists.

Priestley reacted to the initial criticism of his work by trying to repeat his investigations in 1778.

This time he found his experimental results did not, in general, fit with his original hypothesis.

### 2.2.3 Jan Ingen-Housz (1730–1799)

The reason why Priestley and others could not repeat his original experiments successfully became clear in 1779 when Jan Ingen-Housz, a Dutch physician, wrote about certain of his investigations with plants.

*Experiments upon vegetables* by Jan Ingen-Housz (London 1779)

'I observed that plants not only have a faculty to correct bad air in six or ten days, by growing in it, as the experiments of Dr. Priestley indicate, but that they perform this important office in a complete manner in a few hours; that this wonderful operation is by no means owing to the vegetation of the plant, but to the influence of the light of the sun upon the plant.'

Thus, the importance of light in the 'wonderful operation' was first noted and it became obvious that the inconsistency of previous results was due to the varying effects of different light conditions.

Ingen-Housz, in the same series of experiments, formed many more conclusions about the process. Amongst the more important of these were:

'. . . . that this office is not performed by the whole plant, but only by the leaves and the green stalks that support them;'

'. . . . that all plants contaminate the surrounding air by night, and even in the day-time in shaded places.'

So, the importance of the green colour of the vegetation was noted and also the fact that plants had the same effect on air as did animals (in the absence of light).

**SAQ 11** Write down a modified version of equation 2 (page 9) which summarises the conclusions reached by Ingen-Housz about the restoring effects of vegetation on air.

## 2.2.4 Nicholas Theodore de Saussure (1767–1845)

The manner in which the effects of vegetation on air were related to the nutrition of that vegetation was established in 1804 by Nicholas Theodore de Saussure, a Swiss scientist.

Van Helmont's work had indicated that

$$water \longrightarrow vegetable\ matter$$

Priestley and Ingen-Housz had shown that

$$injured\ air \xrightarrow[\text{green vegetation}]{\text{light}} restored\ air$$

At the time of de Saussure's investigations, other scientists had discovered the chemical composition of injured and restored air. De Saussure used this infor-mation and related the 'restoring' process to plant nutrition to show that

$$water\ +\ carbon\ dioxide \xrightarrow[\substack{\text{green} \\ \text{vegetation}}]{\text{light}} \begin{array}{c} vegetable\ matter \\ +\ oxygen \end{array}$$

The following extract from his work demonstrates the need for both water and carbon dioxide (which he called carbonic acid) in the nutrition of green vegetation.

*Chemical investigations of plant growth* by Nicholas Theodore de Saussure (1804).

'The functions of water and gases in the nutrition of plants, the changes that the latter produce in their atmosphere – these are the subjects that I have most investigated. The observations of Priestley, Senebier, and Ingen-Housz have opened the road that I have traversed, but they have not at all attained the goal that I set myself. If, in several instances, imagination has filled the gaps that these observations have left, it has been by conjectures the obscurity and opposition of which have always shown them to be uncertain . . .

The solution of these problems often involves data that we lack completely; exact procedures for the analysis of plants, and a perfect acquaintance with their organisation, are required . . . I attack the problems that can be decided by experiment, and I abandon those that can give rise only to conjectures . . .

One can judge whether the dry or solid structure of plants is increased by the fixation of the constituent principles of water by drying at room temperature a plant similar to, and of the same weight as, that which has been grown in a closed vessel with pure water and oxygen gas. One then remarks whether the plant grown under these conditions has a greater dry weight than it would have had if it had been dried before the experiment, as was the dried plant that serves as a standard of comparison. It is obvious that the two plants must be taken up at the same degree of maturity, from the same soil, and that the weighings must always be made at the same readings of the thermometer and hygrometer.

The numerous experiments that I have made by this procedure have proved to me that plants grown in water alone, in a closed vessel with atmospheric air freed of its carbonic acid gas, do not under these conditions increase the dry weight of their vegetable substance to any appreciable extent. If there is any increase at all, it is by a very small, very limited quantity – one which cannot be further increased by a prolongation of the vegetation . . .

. . . It is very probable that the quantities of hydrogen and oxygen in plants cannot be increased beyond certain limits without correspondingly increasing the amount of their carbon.

Consequently, I have grown plants in a mixture of common air and carbonic acid gas, in order that they might be able to assimilate carbon. In all cases in which the plants have flourished the results were then more pronounced. The plants plainly increased in the weight of their dry vegetable matter, by a quantity larger than that which they would have secured from the elements of the acid gas.'

**6   Summary of discoveries leading to an understanding of photosynthesis**

**SAQ 12** (*a*) How does de Saussure's account show his concern for
(i) investigating very specific problems rather than wide-ranging topics?
(ii) controls in his investigations?
(iii) repeating his investigations to show the consistency of his results?

(*b*) What do you understand by
(i) dry weight?
(ii) fixation – as used in de Saussure's work?

### 2.2.5 Summary of the major discoveries leading to an understanding of photosynthesis

Figure 6 summarises the series of discoveries made by the scientists mentioned in this section.

**SAQ 13** Using the information in figure 6, explain the nature of photosynthesis in one sentence.

We now know that it is green pigment chlorophyll found in plants which is responsible for absorbing the light required in photosynthesis. We also know that the initial products of photosynthesis are sugars.

| Date of publication | Scientist | Discovery | Equation for plant nutrition |
|---|---|---|---|
| 1648 | J. B. van Helmont | All vegetable matter was formed from water | water $\xrightarrow{\text{vegetation}}$ vegetable matter |
| 1772 | J. Priestley | Vegetation could change injured air to restored air | injured air $\xrightarrow{\text{vegetation}}$ restored air |
| 1779 | J. Ingen-Housz | Restoring injured air required light and green vegetation | injured air $\xrightarrow[\text{light}]{\text{green vegetation}}$ restored air |
| 1790 | Various scientists | Injured air was carbon dioxide. Restored air was oxygen | carbon dioxide $\xrightarrow[\text{light}]{\text{green vegetation}}$ oxygen |
| 1804 | N. T. de Saussure | Changing carbon dioxide to oxygen was part of plant nutrition. Minerals were also needed | water + carbon dioxide $\xrightarrow[\text{light + minerals}]{\text{green vegetation}}$ vegetable matter + oxygen |

The following equation is often used to summarise the process of photosynthesis.

$$\text{carbon dioxide} + \text{water} \xrightarrow[\text{light energy}]{\text{chlorophyll}} \text{glucose} + \text{oxygen}$$

$$6CO_2 + 6H_2O \longrightarrow C_6H_{12}O_6 + 6O_2$$

## Practical A: To demonstrate photosynthesis and its requirements

The carbohydrate produced in photosynthesis is stored by many plants as starch. The production of starch by the green parts of plants can therefore be taken as an indication of photosynthetic activity.

You will first carry out the procedure necessary for demonstrating the presence of starch in leaves and then you will use this test to demonstrate:
(i) the need for light in photosynthesis,
(ii) the need for chlorophyll in photosynthesis.

### Materials

*Impatiens balsamina* (busy lizzy) or *Pelargonium* sp. (geranium), beaker, cork-borer (8–10 mm bore), forceps, boiling tube, water-bath, ethanol, iodine in aqueous potassium iodide solution with glass dropper, aluminium foil, thermometer, white tile, light source

### Procedure for demonstrating the presence of starch in leaves

(a) Cut three leaf discs from the leaves available, using a cork-borer.

(b) Immerse the discs in boiling water for 10 s to break down the cell membrane and destroy the enzymes.

(c) Heat a boiling tube of ethanol to 80 °C in a water bath. ☙ If a Bunsen is used to heat the water bath **it must be turned off before the ethanol is placed in the bath.** Transfer the discs to the ethanol until all the pigments have been removed.

(d) Rinse the discs in the warm water to remove the ethanol.

(e) Lay the discs on a tile and test for the presence of starch, using iodine with potassium iodide solution.

### Procedure for obtaining starch-free leaves

(a) Put the plants to be used in a light-proof container for about 48 h.

(b) Test a leaf for the presence of starch. If none is present, carry out the experiment. If starch is still present, return the plant to its light-proof container until it is de-starched.

### Procedure for investigating the need for light in photosynthesis

(a) Attach a strip of aluminium foil across a leaf of a de-starched plant (see figure 7).

**7   The need for light in photosynthesis**

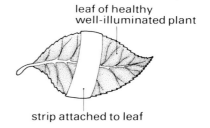

leaf of healthy
well-illuminated plant

strip attached to leaf

(b) Illuminate the whole plant for 4 h.

(c) Test the leaf for the presence of starch and compare the results for those areas having light and those without.

### Procedure for investigating the need for chlorophyll in photosynthesis

(a) Select a de-starched plant (see figure 8).
(b) Record the pattern of chlorophyll distribution by means of a sketch diagram.
(c) Illuminate the plant for 4 h.
(d) Remove one leaf and test for the presence of starch. Compare the results for those areas with chlorophyll and those without.

**A variegated leaf**

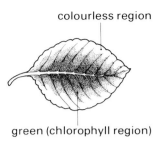

colourless region

green (chlorophyll region)

## Discussion of results

1 What colour did the leaf go in the test for starch?

2 What happens to the starch during the procedure for obtaining starch-free leaves.

3 In investigating the need for light in photosynthesis
a) which parts of the leaf contained starch?
b) what is the most obvious reason for the distribution of starch?
c) suggest at least two alternative explanations.

4 In investigating the need for chlorophyll in photosynthesis
a) describe the distribution of starch in the variegated leaf;
b) suggest a reason for this distribution.

Show this work to your tutor.

### 2.2.6 Summary assignment for section 2.2

Write one or two sentences to describe plant nutrition.

Write out the full equation for photosynthesis in words and in formulae.

Show this work to your tutor.

Self test 2, page 88, covers section 2.2 of this unit.

## 2.3 The effects of internal and external factors on the rate of photosynthesis

Once scientists had discovered the nature of the basic materials for, and the products of, photosynthesis, they turned their attention to the effects of external factors on the rate of photosynthesis. They measured the rate of photosynthesis in terms of the quantity of carbohydrates or oxygen formed.

The rate of photosynthesis is controlled by many external and internal factors. The internal factors include age, chlorophyll type and concentration, enzyme and water content, leaf structure and stomatal aperture. The external factors include carbon dioxide concentration, the wavelength and intensity of light, temperature, wind velocity, water and nutrient supply.

The effect of many of these factors is difficult to determine quantitatively because of the interaction between them. The effects of carbon dioxide concentration, light and temperature can be measured, and this section is largely concerned with these external factors.

### 2.3.1 The effect of carbon dioxide concentration on the rate of photosynthesis

The average carbon dioxide content of the atmosphere is about 0.03–0.04%. An increase in the carbon dioxide content of the atmosphere up to 0.5% results in an increase in photosynthetic rate.

This concentration of carbon dioxide can cause leaf damage, however, and the optimum carbon dioxide concentration seems to be about 0.1% (see figure 9).

9 **The effect of carbon dioxide concentration on the rate of photosynthesis**

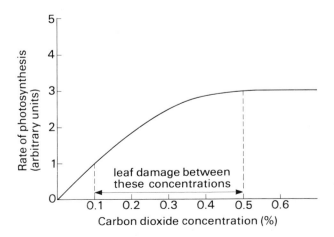

leaf damage between these concentrations

*Rate of photosynthesis (arbitrary units)* vs *Carbon dioxide concentration (%)*

**SAQ 14** (*a*) What makes 0.1% carbon dioxide the optimum in figure 9?
(*b*) Why is there little point in recording rates much beyond 0.5% carbon dioxide concentration?

## 2.3.2 The effect of light intensity on the rate of photosynthesis

In 1905, F. F. Blackman carried out investigations into the effects of light on photosynthesis. He studied the effects of variation in light intensity, or brightness. The results from a typical investigation are shown in figure 10.

**10   The effect of light on the rate of photosynthesis**

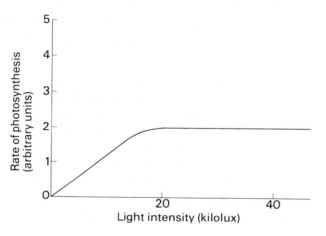

**11   Graph showing the relationship of light intensity and compensation point**

**SAQ 15** What was the effect of increasing the light intensity
(*a*) at relatively low light intensities (less than 20 kilolux)?
(*b*) at higher light intensities (above 20 kilolux)?

## 2.3.3 Light intensity and the compensation point

In total darkness a plant does not photosynthesise, but respires only and thus $CO_2$ is evolved. At low light intensities $CO_2$ is still produced during respiration, but photosynthesis is now occurring so some of the $CO_2$ produced is used up. Therefore, the overall $CO_2$ production falls.

As the light intensity increases, the photosynthetic rate increases. Eventually, a point is reached at which all the $CO_2$ evolved in respiration is used up in photosynthesis, so that is there is no net loss or gain of $CO_2$ by the plant. This is known as the **compensation point**.

As light intensity increases beyond the compensation point, $CO_2$ is taken up from the atmosphere and oxygen is evolved. Once the compensation point has been passed, sugars are produced by photosynthesis at a greater rate than they are used up by respiration. Therefore, a surplus of sugars is produced which can be used by the plant. This information is summarised in figure 11.

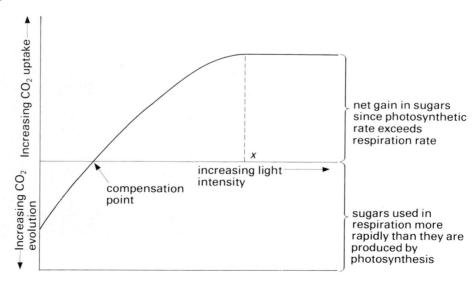

**SAQ 16** Suggest why the photosynthetic rate levels off at light intensity *x*.

Different plants reach their compensation point at different light intensities. Plants which normally grow in sunny environments reach their compensation point at a higher light intensity than plants which normally grow in shady places. For instance, woodland trees normally have a higher compensation point than their seedlings. Such plants are referred to as **sun plants** and **shade plants** respectively.

In shady environments, plants with a low compensation point will use the light more efficiently and will therefore be at an advantage compared with sun plants. Sun plants, however, are much more efficient than shade plants at using high intensity light. Therefore, in sunny environments they will out-compete any shade plants.

Examine figure 12. **A** and **B** represent two woodland plants. One is from the tree layer, the other is from the herb layer beneath the trees.

**12   Carbon dioxide exchange for two plants at varying light intensities**

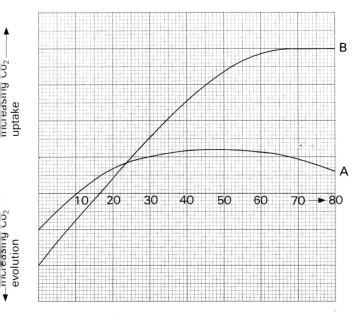

**SAQ 17** For both curves, state
(*a*) whether it represents a sun plant or a shade plant;
(*b*) where in the wood you would expect it to grow;
(*c*) its compensation point.

### 2.3.4 Adaptation to the intensity of light

Plants have adaptations which enable them to regulate the amount of light absorbed by individual leaves and the plant as a whole.

Usually, about 80% of light falling on a leaf is absorbed, with 10% being reflected by the epidermal surface and 10% transmitted through the leaf. The way in which a plant can control the amount of light falling on the leaf depends on the total surface area of the leaves, and the arrangement and orientation of the leaves.

The manner in which leaves of a plant are arranged in relation to each other will also affect the efficiency of the whole plant in photosynthesis. If the upper leaves receive most of the light and the lower leaves are very shaded, this could be much less efficient than all the leaves getting an equal proportion. The alternate arrangement of leaves in the busy lizzy (*Impatiens balsamina*) in figure 13 illustrates one way in which maximum light absorption is achieved.

**13   Arrangement of leaves in a plant**

The angle at which a leaf lies in relation to the rays of sunlight will affect the amount of light falling on the leaf. Many leaves change position during the day in relation to the sun's movements.

## Practical B: The production of oxygen as an indicator of photosynthetic rate

Oxygen is evolved during photosynthesis. Its production can be used as an indication of photosynthesis and as a measure of photosynthetic rate.

### Materials

Young shoot of *Elodea canadensis* in pond water. 400 cm$^3$ beaker, glass funnel to fit into beaker, boiling tube, plasticine, bench lamp with 100 W bulb or equivalent strip lighting, about 10 g potassium hydrogen carbonate, 250 cm$^3$ beaker, stop watch, thermometer

### Procedure

(a) Set up the apparatus as shown in figure 14.

**14   Collecting oxygen from *Elodea***

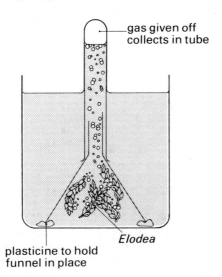

gas given off collects in tube

plasticine to hold funnel in place

*Elodea*

(b) It is important that the test-tube is full of water at the beginning of the demonstration.

(c) Leave the apparatus in artificial light for several days and go onto (d). When you have collected a test-tube of gas, test it for its oxygen content by lowering a glowing splint into it. An increase in intensity of the glow or relighting of the splint indicates the presence of more oxygen than in normal air.

(d) Select a piece of *Elodea* which has a steady stream of bubbles coming from it. Place it in a 250 cm$^3$ beaker of pond water. If bubble production is too slow, addition of a little potassium hydrogen carbonate should increase the rate.

(e) Darken the room and position the lamp so it shines on the *Elodea* and is 10 cm away from it.

(f) Measure the temperature of the water in the beaker and check that it remains constant throughout your investigation.

(g) Allow a settling down period of about 5 min, then count the number of bubbles evolved over a period of 5–10 min depending on the rate of bubbling.

(h) Increase the distance between the light source and the *Elodea* to reduce the intensity of light on the plant and repeat procedure (g). Repeat for about five different distances.

### Recording and discussion of results

The intensity of light is inversely proportional to the square of the distance between the light source and the point it falls on.

$$\text{intensity} \propto \frac{1}{(\text{distance})^2}$$

Thus, if the distance is doubled, the light intensity is reduced to $(\frac{1}{2})^2$ or $\frac{1}{4}$ of its initial value.

Record your results in a table.

Plot a graph of volume of gas collected (number of bubbles) on the vertical axis against light intensity ($1/d^2$) on the horizontal axis (see figure 15).

**15 Volume of gas collected and light intensity**

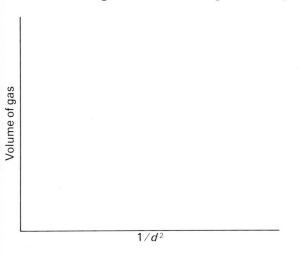

**Questions**

**1** What can you say about the nature of the gas evolved by *Elodea*?

**2** What is the effect of light intensity on the rate of photosynthesis?

**3** How might heating effects influence the results? How could such effects be avoided?

**4** Why should the addition of hydrogen carbonate ions increase the rate of bubbling?

**5** How would you modify the above technique to investigate the effect of light quality (wavelength) on the rate of photosynthesis?

Show this work to your tutor.

---

**2.3.5 The effect of wavelength of light on the rate of photosynthesis**

Investigations into the effect of the wavelength of light on the rate of photosynthesis were carried out by a German scientist, T. W. Engelmann, in 1881. He used the distribution of the motile, oxygen-requiring bacterium *Pseudomonas* as an indication of the rate of photosynthesis in different regions of a filament of the green alga, *Spirogyra* (see figure 16).

*SAQ 18* (*a*) Explain why the distribution of the bacteria could be used as an indication of the rate of

**16 The results of Engelmann's investigation**

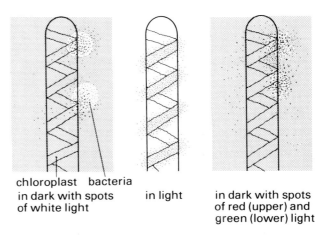

chloroplast  bacteria
in dark with spots    in light    in dark with spots
of white light                      of red (upper) and
                                    green (lower) light

photosynthesis in different regions of the *Spirogyra*.
(*b*) *Spirogyra* has an easily observed, spiral chloroplast. What evidence is there in figure 16 for a relationship between chloroplasts and photosynthesis?
(*c*) Is there any evidence in figure 16 for red light being more effective than green light in photosynthesis?

In 1882, Engelmann was able to observe the effects of a range of wavelengths from the visible spectrum of light on the rate of photosynthesis. He used *Cladophora*, another filamentous green alga, which has cells uniformly filled with chloroplasts. A portion of the *Cladophora* was illuminated with a spectrum of light and, once again, the movements of bacteria were observed to assess the rate of photosynthesis (see figure 17).

**17 *Cladophora* illuminated with a spectrum of light**

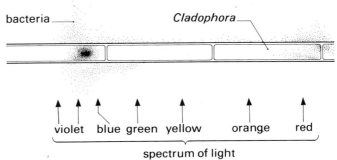

bacteria                          *Cladophora*

violet  blue  green  yellow      orange    red

spectrum of light

*SAQ 19* What evidence is there that the rate of photosynthesis varies with the wavelength of light available?

## 2.3.6 Action and absorption spectra

The type of investigation carried out by Engelmann led to the determination of the **action spectrum** which shows the relative amount of photosynthesis going on at different wavelengths of light (see figure 18).

**18   Action and absorption spectra**

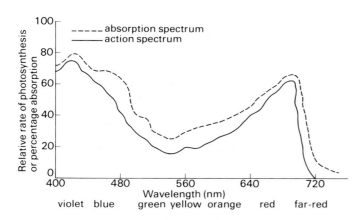

The graph (figure 18) also shows the **absorption spectrum** for a chloroplast extract, that is, the extent to which the green pigment extracted from chloroplasts absorbs different wavelengths of light. Engelmann's investigations had indicated a possible relationship between chloroplasts and photosynthesis. Comparison of the absorption spectrum of a chloroplast extract and the action spectrum for photosynthesis provided evidence for the role of this extract in photosynthesis.

*SAQ 20* Explain how a comparison between the absorption and action spectra in figure 18 provides evidence for the role of chloroplast pigments in photosynthesis.

---

## Practical C: Thin-layer chromatography of leaf pigments

---

This investigation involves extraction and separation of the photosynthetic pigments from leaves using a technique known as thin-layer chromatography. You will then examine the nature and function of these pigments.

**Materials**

Approximately 2 g leaves (nettles or spinach), a little pure sand, pestle and mortar, 4 cm³ 90% acetone (propanone), boiling tube with stopper, test-tube rack, 5 cm³ pipette, ❀ 3 cm³ petroleum ether (light petroleum), glass rod, dropper pipette, specimen tube 7.5 × 2.5 cm and cork with groove to fit chromatography plate (see figure 19), test-tube, thin-layer chromatography plate pre-coated with silica gel cut to about 7 × 2 cm, fine glass capillary tube for spotting, 5 cm³ chromatography solvent (100 parts petroleum ether (boiling point 60–80 °C): 20 parts pure acetone), pin, clear Fablon 3 × 8 cm

**19   Apparatus for thin-layer chromatography**

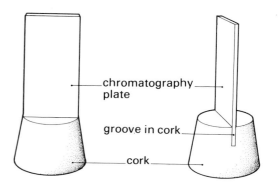

**Procedure**

(*a*) Grind the leaves with a little sand, and a little water, in a pestle and mortar.

(*b*) Place the macerate in a boiling tube containing 4 cm³ acetone. Stopper the tube and shake carefully for 10 s.

(*c*) Stand the tube in a rack for 10 min.

(*d*) Add 4 cm³ water and shake.

(*e*) Add 3 cm³ petroleum ether. Stir with a glass rod. Shake for 5 s. Allow to stand. The solvents will separate.

(*f*) Remove the top layer of chloroplast extract into a test-tube.

(*g*) Pour a few cm³ of chromatography solvent into the specimen tube so that when the cork with the fitted plate is inserted, the level of the solvent is higher than the lower edge of the slide but does not rise to a height of 0.5 cm above this lower edge. Cork the tube and leave for 10 min to allow the atmosphere inside to become saturated.

(h) Use the fine glass capillary tube to place a drop of the chloroplast extract on to a point 1 cm from the lower edge of the chromatography plate. Let the drop dry thoroughly and place a second drop on the same spot. Repeat 10 times. It is important to get a concentrated spot with a diameter no greater than 0.5 cm.

(i) Fit the plate into the cork and place in the specimen tube, ensuring that the lower end of the plate comes into contact with the solvent, but that the solvent does not come into contact with the sample spot.

(j) Leave for 2–25 min until the solvent front has almost reached the top of the plate.

(k) Remove the plate. Mark the solvent front with a pin and leave to dry.

(l) Examine your chromatogram carefully. This may be done in ultraviolet light.

(m) To transfer the chromatogram to your practical file, peel the paper backing from a small piece of clear Fablon and place it sticky side upwards on a piece of glass. Place your chromatography plate, chromatogram downwards, on to the Fablon. Run the forefinger over the back of the plate gently at first and then fairly firmly. Turn the plate over and press with the forefinger for 15 s over the whole surface. Gently lift the Fablon at the corner and peel off the plate. The chromatogram will have adhered to the underside of the Fablon. This may then be inserted into your file. You should draw around the spots and note the colours since they will eventually fade.

## Explanation of chromatography

As the solvent ascends the chromatography plate, it carries the chloroplast pigments with it. Different pigments are carried up at different rates. Thus, a mixture of pigments can be separated from each

other and identified by their different colours and positions.

The relative distances which the different substances move depends on their relative solubility in the liquid component of the thin layer on the chromatography plate. The varying extents to which the different substances are adsorbed onto the surface of the chromatogram is an additional factor in causing the separation.

Under the controlled conditions, the extent to which a particular substance moves up the chromatogram relative to the solvent front is constant. This value is known as the **Rf value**. It can be useful for identifying unknown components in mixtures.

$$\text{Rf value} = \frac{\text{distance moved by substance}}{\text{distance moved by solvent}}$$

## Discussion of results

**1** From your chromatogram, how many different pigments are found in chloroplasts?

**2** For each pigment (a) note its colour and (b) calculate its Rf value.

**3** Compare your results with the chromatogram in figure 20. Suggest reasons for any differences between your results and figure 20.

**20   A pigment chromatogram**

| | | | Rf value |
|---|---|---|---|
| | carotene | (yellow) | 0.95 |
| | phaeophytin | (yellow-grey) | 0.83 |
| | xanthophyll | (yellow-brown) | 0.71 |
| | chlorophyll a | (blue-green) | 0.65 |
| | chlorophyll b | (green) | 0.45 |

Show this work to your tutor.

## Practical D: The absorption of light by chlorophyll

A spectrum showing the different wavelengths of light can be produced using a spectroscope. If a solution of chloroplast pigments is placed in the pathway of the spectrum, it is possible to see what effect these pigments have on light.

### Materials

Spectroscope, chloroplast extract in a flat-sided cell, white light source (clear filament bulb)

### 21   A spectroscope

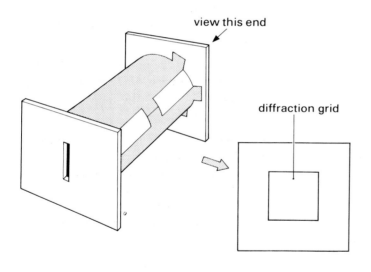

### Procedure

(*a*) A bright light is observed through the spectroscope and the coloured spectrum will be clearly visible within it.

(*b*) A chlorophyll extract is now placed in the beam of light and those colours not absorbed by the chlorophyll extract will still be visible. Note your results.

### Discussion of results

**1** Which colours were transmitted (passed through) the chlorophyll?

**2** Which colours were absorbed?

**3** Which colours have potential for use in photosynthesis by the plant?

Show this work to your tutor.

### 2.3.7 Adaptation to the wavelength of light

From practical C, you learnt that the chloroplasts contain a number of photosynthetic pigments. These are responsible for absorbing light energy. Chlorophyll *a* absorbs red and violet light most strongly. Orange, yellow, blue and indigo are also absorbed. The different accessory pigments absorb different wavelengths. Very little green light is absorbed. Instead, it is reflected from the leaf. This is why the leaf appears green.

In the plant, the light absorbed by the pigments is used in photosynthesis. If a suspension of pigments is prepared in a test-tube, the absorbed light cannot be used and some of it is re-transmitted giving a red colour to the solution. This is known as fluorescence.

Chlorophyll *a* is the pigment directly involved in the photochemical reactions of photosynthesis. The other pigments act as accessory pigments, by capturing the light energy and then passing it on to chlorophyll *a*.

*SAQ 21* Name two accessory pigments.

Accessory pigments are important because they enable a plant to make best use of the wavelengths of light in its own particular environment.

In a woodland, light filtering through the trees to reach the shade plants of the undergrowth will have less blue and red light since this is removed by chlorophyll *a* in the sun plants. Shade plants have a higher concentration of chlorophyll *b* whose absorption peaks are more towards the centre of the visible spectrum (see figure 22). Thus, they make maximum use of that light which reaches them.

**22   Absorption spectra of chlorophyll *a* and two accessory pigments**

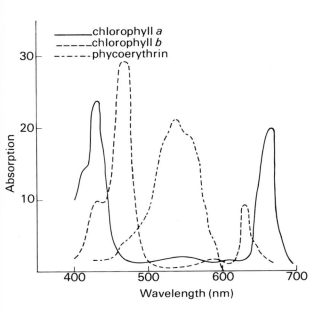

**23   Rate of photosynthesis and the interaction of factors**

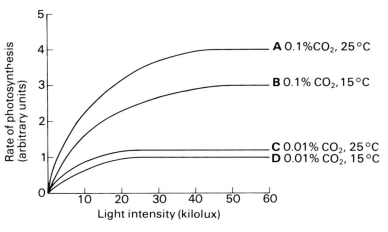

## 2.3.8 The interaction of factors affecting the rate of photosynthesis

Only under laboratory conditions are plants likely to be subjected to variation of just one environmental factor. Normally, more than one will vary and the resultant rate of photosynthesis will be due to the interaction of these factors. The effect of light intensity on the rate of photosynthesis of a suspension of the unicellular green alga *Chlorella* at two different temperatures and carbon dioxide concentrations is shown in figure 23.

Curve **D** shows the rate of photosynthesis (at 0.01% $CO_2$ concentration and 15 °C temperature) increasing from 0 kilolux light intensity to a maximum at about 20 kilolux. Between 20 and 40 kilolux, there is no increase in the rate of photosynthesis – the reaction is said to be limited. One of the factors which affect the rate of photosynthesis must be at its maximum usage. There are a number of possibilities.
(*a*) The reaction can go no faster under any conditions.

(*b*) The reaction is going as quickly as possible at that temperature.
(*c*) The light energy available at 20 kilolux is the maximum which can be used.
(*d*) The reaction is going as quickly as possible at that $CO_2$ concentration.

*SAQ 22* Consider each of the possibilities (*a*) to (*d*) in turn and explain how the information provided by curves **A, B, C** and **D** in figure 23 supports or contradicts these possibilities.

Photosynthetic rate depends on a number of external factors especially light intensity, $CO_2$ concentration and temperature. For maximum rates to be achieved, **all** factors must be above an optimum level. If any one factor is below its optimal level, the whole process is slowed down. This factor is then said to be the **limiting factor** of the process.

Any process which is dependent on a number of factors may be slowed down by lack of just one of these factors. This phenomenon is known as the **law of limiting factors.**

### 2.3.9 Extension: Interacting factors and the mechanism of photosynthesis

In 1905, Blackman also investigated the effects of varying both temperatures and light on the rate of photosynthesis (see figure 24).

**24  The effects of temperature and light together on the rate of photosynthesis**

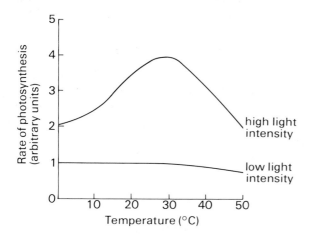

**25  The effect of temperature on the rate of chemical reactions at low and high light intensity**

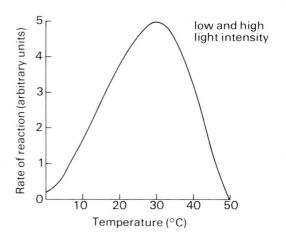

*SAQ 23* What was the effect on the rate of photosynthesis of
(*a*) increasing the temperature between 30 °C and 40 °C;
(*b*) increasing the temperature between 0 °C and 30 °C at (i) high light intensity, (ii) low light intensity?

The results from these investigations led Blackman to propose that there were at least two reactions involved in photosynthesis – one that depended on light energy and one which was a chemical reaction independent of light energy.

In order to understand Blackman's reasoning, consider the effects of light and temperature on two different kinds of reactions.

*(i) Chemical reactions independent of light energy*

Chemical reactions proceed faster as the temperature increases. In biological systems this happens up to about 40–45 °C when the effect of enzyme denaturation begins to influence rates. Such reactions are not affected by light energy changes (see figure 25).

*(ii) Light reactions driven only by light energy*

Light reactions are not affected by temperature. They will proceed faster only if the light intensity is increased (see figure 26).

**26  The effect of temperature on the rate of photoreactions at low and high light intensity**

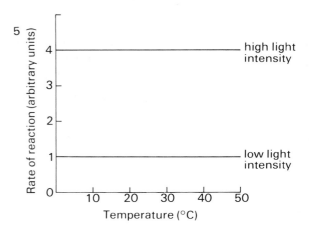

Blackman's investigation produced results which had the characteristics of a light-dependent reaction at low light intensities and a chemical reaction independent of light at high light intensities. He therefore suggested this was due to the presence in photosynthesis of an initial light-dependent reaction followed by a chemical light-independent reaction. These have become known as the light and dark reactions of photosynthesis – an unfortunate name in the case of the dark reaction because it implies a necessity for darkness rather than the actual case of an independence from light.

## Practical E: Investigations using carbon dioxide as an indicator of photosynthesis

Carbon dioxide is taken up as one of the raw materials for photosynthesis. This uptake can therefore be used as an indication of photosynthetic activity.

In this practical you will use an indicator solution to detect the pH changes in a solution. Carbon dioxide in solution forms carbonic acid and if the carbon dioxide is removed from a solution it therefore raises its pH.

The indicator you have available will change from (i) green to blue (if bromothymol blue is used), or (ii) orange to purple (if bicarbonate indicator is used), as it becomes more alkaline. The above colour changes will therefore be indicators of carbon dioxide uptake from the solution.

### Materials

Indicator solution (bromothymol blue or bicarbonate) at neutral pH, 8 test-tubes (25 ml) and rubber bungs, aluminium foil, muslin, cotton wool, cork-borer (size 8–12), strip lighting or lamp (60 W) and beaker of water as heat filter, test-tube rack, syringe (2 cm³), forceps, ruler, marking pen, live plant material (e.g. *Impatiens, Pelargonium, Elodea*)

### Procedure

(*a*) It is important to avoid any contamination which would influence the performance of the indicator. Care must be taken to ensure that:

(i) all test-tubes are thoroughly clean (they should have been well washed in hot water and rinsed twice, first in distilled water and then with indicator solution);
(ii) the plant material is washed in distilled water and rinsed with indicator solution;
(iii) you avoid breathing into the test-tubes.

(*b*) Set up a test-tube with 2 cm³ of indicator solution in it.

(*c*) Close the tube with a rubber bung.

(*d*) Leave the tubes for 10 min to allow the indicator solution to equilibrate with the air in the tube. The indicator should remain at its neutral colouration (green for bromothymol blue, orange for bicarbonate indicator).

(*e*) Place a sprig of plant material in the indicator and replace the bung (see figure 27). Illuminate the tube for one hour with strip lighting, or a 60 W lamp placed 20 cm away with a beaker of water as a heat filter between the tube and the lamp.

### 27   Uptake of carbon dioxide by a leaf

(*f*) Observe any colour changes in the indicator solution.

### Discussion of results

**1** Describe the colour change in the indicator.

**2** What does this indicate about $CO_2$ uptake or output?

**3** What processes are responsible for these changes?

## Further investigations

Using carbon dioxide as an indicator of photosynthesis, design and carry out further investigations to answer the following questions.

What is the effect on the rate of photosynthesis (as measured by carbon dioxide uptake) of
(*a*) having different intensities of light?
(*b*) different-sized areas of leaf?
(*c*) the same-sized areas of leaf from different leaves of the same plant?
(*d*) the same-sized areas of leaf from different species of plant?

**28   This set-up could be used for further investigation (*a*) and modified for (*b*), (*c*) and (*d*)**

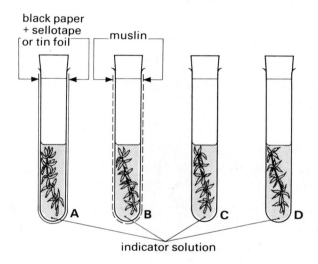

## Results

Your results should be recorded and interpreted to answer the questions above.

Write an essay plan for the following past examination question. Discuss it with your tutor. Then write the essay and hand it to your tutor for marking.

Write an account of the ways in which external factors may influence the rate of photosynthesis in a green plant. Describe carefully how you would measure:
(*a*) the rate of photosynthesis, and
(*b*) the effect of one external factor on this rate.
(London, January 76)

24

Show this work to your tutor.

---

### 2.3.10 Summary assignment for section 2.3

**1** List six internal, and seven external, factors which affect the photosynthetic rate.

**2** Sketch graphs to show the following:
(*a*) the effects of increasing $CO_2$ concentration on photosynthetic rate (assuming other factors are constant);
(*b*) the effects of increasing light intensity on photosynthetic rate (assuming other factors are constant);
(*c*) the action spectrum for photosynthesis;
(*d*) the absorption spectrum for chlorophyll *a*;
(*e*) the relationship between photosynthetic rate, light intensity and compensation point in sun and shade plants.

**3** Name five pigments found in green plants.

**4** State the similarities and differences between chlorophyll *a* and accessory pigments.

**5** Sketch the graph in figure 23 and state what it shows of:
(*a*) the effects of increasing light intensity on photosynthetic rate;
(*b*) the effects of increasing $CO_2$ concentration on photosynthetic rate;
(*c*) the effects of temperature on photosynthetic rate.

**6** State the law of limiting factors.

**7** Summarise Blackman's arguments for suggesting photosynthesis occurs in two stages.

Show this work to your tutor.

Self test 3, page 88, covers section 2.3 of this unit.

## 2.4 The mechanism of photosynthesis

Two different phases are believed to be involved in the mechanism of photosynthesis:
(i) a light-dependent phase, in which light energy is trapped and converted into chemical energy;
(ii) a light-independent phase, in which this chemical energy is made available for the reduction of carbon dioxide to carbohydrate.

This view of the mechanism of photosynthesis is summarised in figure 29. Here, the rectangle represents photosynthesis and the two different reactions occupy different spaces within the rectangle.

**29    The raw materials and products of photosynthesis**

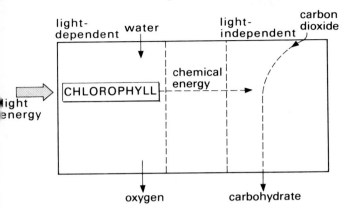

## 2.4.1 The Hill reaction

The light-dependent phase involves the use of light energy to produce energy-transferring molecules of adenosine triphosphate (or ATP) and a reducing agent, known as reduced nicotinamide adenine dinucleotide phosphate (or NADPH$_2$). During this process oxygen is evolved.

In 1937, Robert Hill, a British scientist, developed a technique for isolating and working with chloroplasts. He showed that illuminated isolated chloroplasts could produce oxygen and a reducing agent. This became known as the Hill reaction.

$$4Fe^{3+} + 2H_2O \xrightarrow[\text{chloroplasts}]{\text{light}} 4Fe^{2+} + 4H^+ + O_2$$

## Practical F: Demonstrating the Hill reaction

The Hill reaction involves the production of a reducing agent in the presence of light by plant material. This may be demonstrated by following the reduction of iron(III) chloride in a chloroplast suspension to iron(II) chloride using iron(II)-sensitive reagent strips.

## Materials

10 g sucrose, ☠ 0.25 g iron(III) chloride (POISONOUS!), 100 cm$^3$ water, 10 g spinach, tissue paper, muslin, liquidiser, iron(II)-sensitive reagent strips, hydrogen peroxide, fine dropper, distilled water, boiling tubes, measuring cylinder

## Procedure

(*a*) Dissolve 10 g sucrose and 0.25 g of iron(III) chloride in 100 cm$^3$ water. The sucrose is present to prevent the chloroplasts absorbing too much water causing them to burst.

(*b*) Pour the solution into a liquidiser together with 10 g of fresh spinach.

(*c*) Blend for 30 s.

(*d*) Filter through fine muslin.

(*e*) Test filtrate for Fe$^{2+}$ with the Fe$^{2+}$-sensitive reagent strips. If any Fe$^{2+}$ is present, add very small drops of hydrogen peroxide until it is removed.

(*f*) Divide the filtrate into two 50 cm$^3$ samples in boiling tubes.

(*g*) Place one sample in the dark and the other 10 cm from a 60 W light source.

(*h*) Test both samples with reagent strips at 10 min intervals and record your results in a table indicating the concentration of Fe$^{2+}$ ions in solution in parts per million. The chlorophyll may colour the strip and must be wiped off with a clean tissue.

## Discussion of results

**1** Do your results indicate that a reducing agent can be produced by plant material?

**2** Does light appear to be essential for this process?

**3** What is the significance of this process being carried out by 'blended' plant material?

**4** What was the point of procedure (*e*)?

Show this work to your tutor.

### 2.4.2 Arnon's contribution to our knowledge of photosynthesis

The role of light in photosynthesis was further investigated by Daniel Arnon and his associates in America. They improved on Hill's technique for isolating chloroplasts by managing to get chloroplasts which could carry out the whole photosynthetic process, including the conversion of carbon dioxide into carbohydrates in the light.

Daniel Arnon's own account of his work may be read in *The Role of Light in Photosynthesis, Scientific American*, November 1960. Use the QS3R method of reading to help you read and understand this article. (This method is described in the unit *Inquiry and investigation in biology* in this series.)

### 2.4.3 The trapping and conversion of light energy

The trapping of light energy is carried out by the pigment molecules present in green plants. The first part of the reaction occurs when a unit of light, known as a photon, is absorbed by a pigment molecule. The energy may simply be re-emitted or converted to heat. Alternatively, it may cause an electron from the pigment molecule to be 'excited' or raised to a higher energy level.

The higher energy electrons are in an unstable state. These electrons are therefore subsequently released from the pigment molecule and pass along a series of molecules known as **electron carrier molecules**. The movement of electrons from molecule to molecule is known as **electron flow**.
During the passage of electrons from molecule to molecule, energy is released in stages. At some of these stages, the energy release is coupled to the production of ATP and reduced NADP (or NADPH$_2$) (see figure 30).

You have already come across ATP as an energy transferring molecule. NADP is an important molecule associated with photosynthesis. It is able to pick up or donate hydrogen ions. It is therefore known as a hydrogen carrier molecule.

$$NADP + 2H^+ + 2e^- \longrightarrow NADPH_2$$

**30   A summary of energy transfer in the light-dependent phase of photosynthesis**

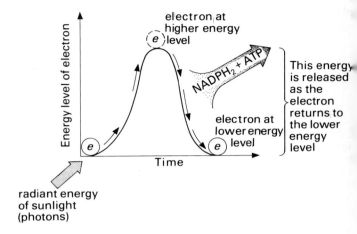

*SAQ 24* What two important substances are produced during the Hill reaction?

*SAQ 25* What did Arnon and his colleagues show?

*SAQ 26* (*a*) What three things could happen to light energy absorbed by chlorophyll molecules?
(*b*) Which of these alternatives is involved in photosynthesis?

*SAQ 27* High-energy electrons pass along a series of electron carrier molecules. What happens to some of the energy during this process?

### 2.4.4 The reactions of the light-dependent phase

There are two major, alternative sets of reactions in the light-dependent phase of photosynthesis. The first, known as **non-cyclic electron flow** as it is a linear set of reactions, occurs when excited electrons pass from the chlorophyll molecule to NADP which is subsequently reduced by these electrons and by hydrogen ions. A supply of hydrogen ions and electrons is made available from water molecules to replenish those used in these reactions.

ATP is synthesised in these reactions, as shown in figure 31. ATP formation in this way is referred to as **non-cyclic photophosphorylation**.

The second set of reactions is cyclic, that is the electrons are returned to the chlorophyll molecule. ATP is produced here also, but water is not involved and no $NADPH_2$ is formed (see figure 32).

**31   Non-cyclic electron flow and photophosphorylation**

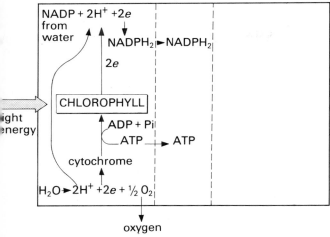

**32   Cyclic electron flow and photophosphorylation**

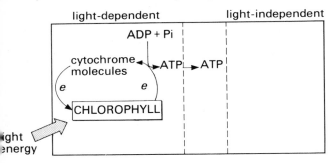

*SAQ 28* During which of the above two types of electron flow sequences is the oxygen which is evolved during photosynthesis produced?

**2.4.5 The role of water**

Water dissociates to form hydroxyl ions ($OH^-$) and hydrogen ions ($H^+$):

$$H_2O \rightleftharpoons H^+ + OH^-$$

The hydroxyl ions may further dissociate to produce electrons and oxygen:

$$2OH^- \rightleftharpoons H_2O + \tfrac{1}{2}O_2 + 2e$$

A water oxidase enzyme speeds up this dissociation. Because illuminated pigments are continually picking up the electrons produced by dissociation, more water molecules dissociate to maintain an equilibrium between the molecules and ions.

This process is sometimes referred to as **photolysis**, implying that the water molecules are split due to the effects of light.

*SAQ 29* State two factors, in addition to light, which are necessary for the water molecules to be 'split'.

For each of the following questions, state which answer/answers are correct. More than one alternative may be correct in each case.

*SAQ 30* Cyclic electron flow involves
(*a*) formation of ATP.
(*b*) formation of $NADPH_2$.
(*c*) flow of electrons from chlorophyll via electron carriers back to chlorophyll.
(*d*) flow of electrons via electron carriers from water to chlorophyll to NADP.
(*e*) evolution of oxygen.

*SAQ 31* Non-cyclic electron flow involves
(*a*) formation of ATP.
(*b*) formation of $NADPH_2$.
(*c*) flow of electrons from chlorophyll via electron carriers back to chlorophyll.
(*d*) flow of electrons via electron carriers from water to chlorophyll to NADP.
(*e*) evolution of oxygen.

*SAQ 32* Phosphorylation
(*a*) describes the synthesis of $NADPH_2$ from NADP.
(*b*) describes the synthesis of ATP from ADP and inorganic phosphate.
(*c*) requires a source of energy.
(*d*) releases energy.

*SAQ 33* The splitting of water during photosynthesis
(*a*) occurs as a direct result of light energy splitting water molecules.
(*b*) occurs as a result of the dissociation of water molecules together with the constant removal of the electrons and hydrogen ions so formed.
(*c*) results in the evolution of carbon dioxide.
(*d*) requires oxygen.

**2.4.6 The source of evolved oxygen in photosynthesis**

In 1941, Ruben and others experimented on photosynthesis in the green alga *Chlorella*.

They used a form of oxygen (isotope) which was heavier than normal: the atomic weight was 18 rather than 16. Otherwise the two forms are chemically identical.

When they used $C^{18}O_2$ instead of $C^{16}O_2$, the heavy isotope appeared in the carbohydrate and in water

$$C^{18}O_2 + 2H_2O \xrightarrow[\text{chlorophyll}]{\text{light}} CH_2^{18}O + O_2 + H_2^{18}O$$

When they used $H_2^{18}O$ instead of $H_2^{16}O$, the heavy isotope appeared in the oxygen.

*SAQ 34* From these results, which of the two raw materials in photosynthesis is the source of the oxygen evolved?

### 2.4.7 Extension: The two photosystems of photosynthesis

It has been shown that there are two reactions in photosynthesis, associated with non-cyclic phosphorylation. The two light reactions are associated with two separate groups of special chlorophyll *a* molecules known as photosystem I and photosystem II. The role of the two light reactions is summarised in figure 33. Study it carefully, starting at the bottom left-hand corner.

**33   The role of the two light reactions in photosynthesis**

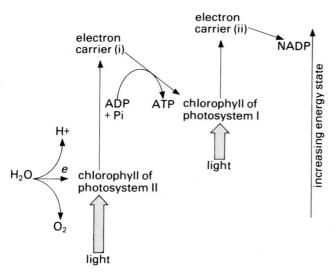

(1) Electrons from water pass to photosystem II where they are raised to a higher energy level by light.

(2) The high energy electrons pass to electron carriers.

(3) Some of their energy is released and this is coupled to the synthesis of ATP.

(4) The electrons then pass to chlorophyll of photosystem I where their energy level is again increased.

(5) The higher energy electrons now pass via electron carriers to NADP.

(6) Notice that molecules towards the top of the diagram have a higher energy state.

(7) N.B. Electron carriers (i) and (ii) are actually made up of several molecules.

*SAQ 35* (*a*) Which photosystem/photosystems are involved in cyclic electron flow?
(*b*) Which photosystem/photosystems are involved in non-cyclic electron flow?

*SAQ 36* What is the action of light on photosystems I and II?

*SAQ 37* Why does the energy of the electrons fall between electron carrier (i) and chlorophyll of photosystem I?

### 2.4.8 Summary of the light-dependent phase of photosynthesis

The two major products of the light-dependent phase of photosynthesis are ATP and $NADPH_2$. These are then used in the light-independent phase to reduce $CO_2$ to carbohydrate.

A summary of the reactions of the light-dependent phase is shown in figure 34.

### 2.4.9 Summary assignment for section 2.4

**1** State Arnon's contribution to our knowledge of photosynthesis.

**2** Using figures 31 and 32, make annotated diagrams to explain cyclic and non-cyclic photophosphorylation. Write the raw materials for each process in one colour and the products in a different colour.

**3** Explain what is meant by 'the Hill reaction'.

**4** Explain the role of chlorophyll, light absorption, electron flow and electron carriers in photosynthesis.

**5** Explain how Ruben and others confirmed that the oxygen evolved in photosynthesis came from water.

Show this work to your tutor.

Self test 4, page 89, covers section 2.4 of this unit.

**34  A summary of the reactions in the light-dependent phase**

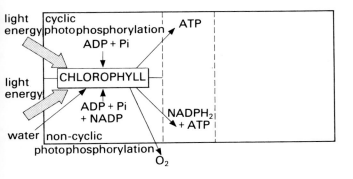

## 2.5 The light-independent reactions of photosynthesis

The sequence of reactions in which carbon dioxide is reduced to carbohydrate in the light-independent phase of photosynthesis was discovered by a group of American scientists led by Melvin Calvin in the late 1940s. This sequence is now known as the **Calvin cycle**.

### 2.5.1 Discovery of the Calvin cycle

Calvin used *Chlorella*, a unicellular green alga, in his investigations and supplied them with radioactive carbon dioxide ($^{14}CO_2$). The carbon atom in the carbon dioxide was $^{14}C$, a radioactive isotope of carbon, rather than $^{12}C$, the non-radioactive isotope which constitutes the major proportion of atmospheric carbon dioxide. Figure 35 shows how the equipment for the investigation was set up.

The *Chlorella* suspension was 'fed' $^{14}CO_2$ and samples were taken from the suspension after time intervals

**35   Calvin's apparatus**

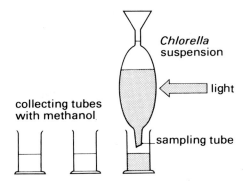

of between 5 and 30 s. These samples were collected in methanol which killed the *Chlorella* immediately, stopping all reactions, including photosynthesis. The chemicals from the *Chlorella* were then analysed and any with $^{14}C$ in them noted. Calvin was thus able to build up a picture of which chemicals were involved in the light-independent phase of photosynthesis and also of the sequence in which they became labelled with $^{14}C$.

The analysis of the chemicals from the *Chlorella* involved two procedures: two-dimensional chromatography and autoradiography.

### 2.5.2 Two-dimensional paper chromatography

The principle of paper chromatography is similar to that of thin-layer chromatography used in practical C. In this process, a sample of the chemical under test is loaded onto a spot near the corner of a piece of chromatography paper.

The paper is then hung in contact with a solvent. This solvent travels up the paper and the component chemicals separate out along the vertical line ($A \rightarrow B$). The paper is then turned through 90° and a second solvent is used to achieve a further separation in the direction ($B \rightarrow C$) (see figure 36).

The second solvent is chosen to produce maximum separation of those molecules which would stay close together with the first solvent. Comparison with the positioning of known chemicals subjected to identical treatment can be used for identification of the chemicals in the original sample. This may involve the calculation of Rf values (see practical C).

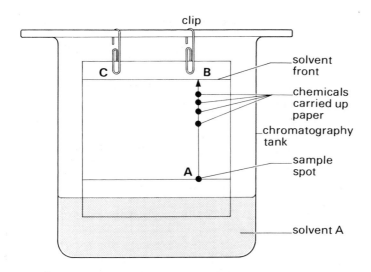

**2.5.3 Autoradiography**

Those chemicals which contain $^{14}CO_2$ can be identified by a further process called autoradiography. The radioactive carbon atoms break down, emitting particles. These particles may be detected by their effect on photographic plates.

The chromatogram is placed face-down on the surface of an X-ray film and left in the dark for a few days. Each spot on the chromatogram which contains a radioactive chemical will cause an adjacent spot on the film to become exposed. This will show as a darkened area after development.

The chemicals found in *Chlorella* which contained radioactive label at various time intervals after feeding it with radioactively labelled carbon dioxide are shown in figure 37. The information is presented here in a more simplified form than that in which it was originally obtained.

From the information given in this table, it is possible to work out part of the sequence of the light-independent reactions precisely. The rest of the sequence requires investigation by means of further experiments.

*SAQ 38* (*a*) Write out that part of the sequence which it is possible to deduce precisely from the information in figure 37.

| Time | Labelled molecules |
|------|--------------------|
| 0 | Carbon dioxide |
| 1 | Phosphoglyceric acid |
| 2 | Phosphoglyceric acid, Triose phosphate |
| 3 | Phosphogylceric acid, Triose phosphate Ribulose phosphate, Glucose |
| 4 | Phosphoglyceric acid, Triose phosphate Ribulose phosphate, Glucose Ribulose diphosphate (ribulose biphospha |

(*b*) Give two possible alternatives for the remainder of the sequence, using only information from figure 37.

**2.5.4 The fixation of $CO_2$**

It was then necessary to discover which molecule $CO_2$ reacted with as it entered the sequence of reactions. This is sometimes known as the fixation of $CO_2$, or **carboxylation**.

Figure 38 shows the levels of PGA (phosphoglyceric acid) and RDP (ribulose diphosphate) in the alga *Scenedesmus* before and after the removal of $CO_2$.

**38   Levels of intermediates of photosynthesis before and after removal of $CO_2$**

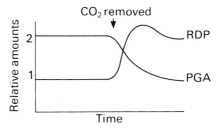

Choose the correct alternative in the following self-assessment questions.

*SAQ 39* It would be expected that the molecule with which $CO_2$ normally combines during photosynthesis would **rise/fall** if $CO_2$ is suddenly removed.

*SAQ 40* The information in figure 38 suggests that this molecule is **PGA/RDP**.

## 2.5.5 The Calvin cycle

It is now known that the light-independent reactions of photosynthesis form a cycle, known as the Calvin cycle. Carbon atoms are continuously introduced into the cycle in the form of $CO_2$ and incorporated into organic carbon molecules.

The molecule with which the $CO_2$ combines (RDP) is regenerated during the cycle to ensure the continuation of the process. During the cycle, intermediates are tapped off for the synthesis of products required by the plant.

The Calvin cycle is outlined in figure 39.

*SAQ 41* Which two reactions in the cycle are dependent on products of the light reaction of photosynthesis?

## 2.5.6 Extension: The four phases of the Calvin cycle

Another way of looking at the Calvin cycle is to

### 39  Outline of the Calvin cycle

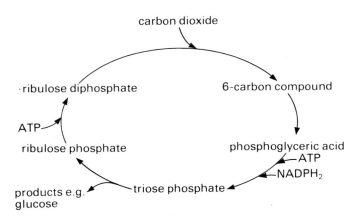

separate it into four phases: carboxylation, reduction, regeneration and product synthesis.

These phases are demonstrated in figure 40, showing the fate of three molecules of RDP during the course of the cycle.

### 40  The four phases of the Calvin cycle

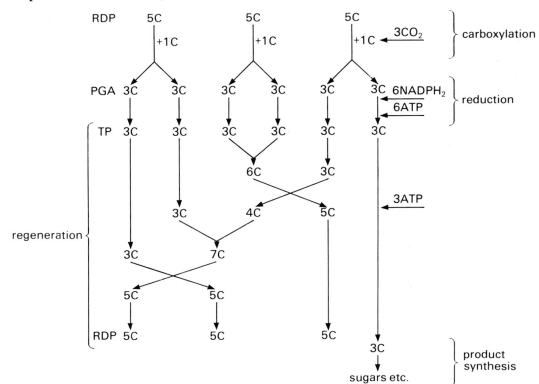

## 41   The products of the Calvin cycle

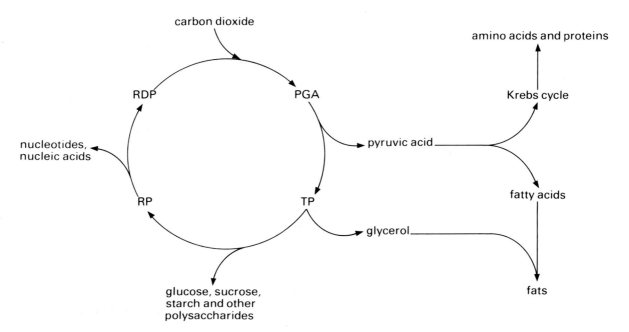

At the start of the diagram, three molecules of the 5-carbon sugar RDP, and three molecules of $CO_2$, enter the cycle.

*SAQ 42* What has happened to these carbon atoms at the end of the diagram?

It is now believed that many organic compounds can be formed directly from the Calvin cycle. Figure 41 illustrates this.

*SAQ 43* Describe the shortest route a carbon atom could follow from carbon dioxide to becoming part of (*a*)a fat, (*b*) an amino acid, (*c*) a polysaccharide.

## 42   Overall summary for the reactions of photosynthesis

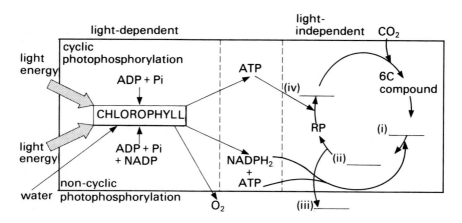

## 2.5.7 An alternative method of carbon fixation

An alternative method of carbon fixation has been demonstrated in certain tropical plants, such as sugarcane and maize.

Carbon dioxide initially becomes fixed into a 4-carbon acid instead of the 3-carbon PGA. The 4-carbon acid is subsequently converted to PGA and enters the Calvin cycle.

Plants which carry out this additional fixation method are called $C_4$ plants to distinguish them from the $C_3$ plants.

*SAQ 44* Complete figure 42 to act as an overall summary for the reactions of photosynthesis.

Write an essay plan to answer the following past examination question. Work out a mark scheme (total 20 marks) to go with the plan. Discuss your plan with fellow students. Write the essay and hand it in for marking.

Describe how the sun's energy is used in food manufacture in a green plant and how the flowering plant uses this food after it has been formed. (London, June 1978).

## 2.5.8 References for sections 2.2–2.5

*Photosynthesis* by G. E. Fogg.
*Photosynthesis* by D. O. Hall & K. K. Rao, Studies in Biology No. 37.
*The Path of Carbon in Photosynthesis* by J. A. Bassham.
*The Role of Light in Photosynthesis* by D. I. Arnon.

The last two references are included in Readings from Scientific American, *The Living Cell* by D. Kennedy.

## 2.5.9 Summary assignment for section 2.5

**1** Explain Calvin's contribution to photosynthesis including reference to the following.
*Chlorella*, $^{14}C$, chromatography, autoradiography

**2** Explain how the initial fixation of $CO_2$ occurs in photosynthesis.

**3** Make a summary diagram of the Calvin cycle based on figures 39, 40, and 41. In your diagram show
(*a*) the important intermediates of the cycle,
(*b*) the phases of carboxylation, reduction and regeneration,
(*c*) the synthesis of carbohydrates, proteins and fats,
(*d*) the introduction of materials from the light-dependent phase,
(*e*) the number of carbon atoms in each molecule in the cycle.

**4** What is a $C_4$ plant?

Show this work to your tutor.

Self test 5, page 90, covers section 2.5 of this unit.

## 2.6 Autotrophic bacteria

Autotrophic bacteria may be divided into two groups, photosynthetic or chemosynthetic. The chemistry of the photosynthetic process in the former group differs in detail from that of other autotrophic plants.

### 2.6.1 Photosynthetic bacteria

These are able to absorb the sun's energy by means of a pigment called bacteriochlorophyll, which is related to chlorophyll. The energy is then made available for reducing carbon dioxide to carbohydrate. Bacteria differ from green plants in their source of hydrogen, which comes from substances other than water. The equation for photosynthesis in sulphur bacteria, for example, is

$$6CO_2 + 12H_2S \xrightarrow[\text{bacteriochlorophyll}]{\text{light}} C_6H_{12}O_6 + 12S + 6H_2O$$

*SAQ 45* Where would you expect to find sulphur bacteria living?

### 2.6.2 Chemosynthetic bacteria

The chemosynthetic bacteria synthesise organic from inorganic materials without using light energy or energy from the oxidation of organic compounds. Instead, they use energy from the oxidation of various inorganic chemicals such as ferrous salts, hydrogen sulphide, hydrogen and ammonia.

The soil bacterium *Nitrosomonas* obtains energy by the conversion of ammonium compounds in the soil to nitrites.

$$(NH_4)_2CO_3 + 3O_2 \longrightarrow 2HNO_2 + CO_2 + 3H_2O + energy$$

The oxidation of ammonia (which is released from the bodies of dead animals and plants by saprophytic bacteria and fungi) is particularly important as it results in the recycling of nitrogen.

*SAQ 46* Which of the statements (*a*)–(*e*) are true for photosynthetic bacteria and chemosynthetic bacteria?
(*a*) Obtain energy from the oxidation of inorganic molecules such as ammonium salts.
(*b*) Obtain energy from sunlight.
(*c*) Contain photosynthetic pigments.
(*d*) Are always green.
(*e*) Are autotrophic.

### 2.6.3 Summary assignment for section 2.6

1 Name the process whereby the sulphur bacteria obtain their nutrients.

2 Outline the role of hydrogen sulphide and bacteriochlorophyll in this process.

3 (*a*) To which nutritional group does *Nitrosomonas* belong?
(*b*) Outline the process by which it obtains its energy.
(*c*) Of what ecological process is this part?

Show this work in to your tutor.

Self test 6, page 90, covers section 2.6 of this unit.

## 2.7 The organ of photosynthesis – the leaf

Photosynthesis takes place in the chloroplasts which are mainly found in plant leaves. This section looks at the structure of leaves. Practical examination of leaves is essential to a good understanding of their structure and function, and the practical exercises in this section should be carried out in the sequence indicated.

### 2.7.1 The surface structure of leaves

Photomicrograph 43 is of the surface of a monocotyledonous leaf. It reveals the surface structure of the epidermal cells and stomata of the leaf.

Each stoma (plural stomata) or pore is bounded by two guard cells. These control the opening and closing of the stoma, and so regulate the rate at which gas exchange occurs between the inside of the leaf and the external environment, and the rate at which water is lost by evaporation from the leaf.

43   Epidermis – surface view. Monocotyledonous leaf ( × 150)

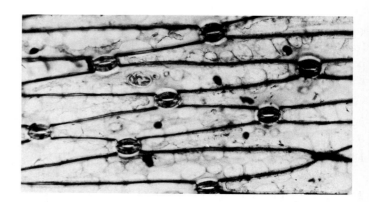

### Practical G: The surface structure of a leaf

#### Materials

*Impatiens* plant (a selection of other leaf types may also be provided), microscope and lamp, slides and cover-slips and soft brush, 5 cm³ acetone, 10 cm × 10 cm acetate sheet (from overhead projector roll), forceps, watchglass

#### Procedure

Examine the surface structure of the leaves provided using the three techniques described below. Both the upper and lower surfaces should be examined. In your observations, note the following points carefully:

(i) the relative sizes of the guard cells and the epider-

mal cells;
(ii) the nature of the cell walls of the guard cells and the epidermal cells;
(iii) the nature of the contents of the guard cells and the epidermal cells.

Make high-power drawings to show a few representative cells from each surface. Label these diagrams fully.

(*a*) Pick a leaf and examine its surface (under the low power of your microscope).

(*b*) Tear the leaf in two and look along the torn edges for an area of exposed epidermis (the outermost cell layer). Mount this in a drop of water on a slide and under a cover-slip. Ensure that the strip of epidermis stays outer-side up and observe the surface structure under low and high power.

(*c*) Cut a rectangle of acetate sheet just a little larger than the leaf. Brush acetone onto the surface of the leaf. Quickly apply the sheet to the treated leaf surface. Press down firmly with the thumb or a glass slide for 15–20 s. Remove the sheet, which will now bear an impression of the leaf surface, and examine it under the low power and the high power of the microscope.

Show this work to your tutor.

### 2.7.2 The internal structure of leaves

Figures 44 and 45 show the internal structure of

**44   Vertical section (VS) of dicotyledonous leaf (× 400)**

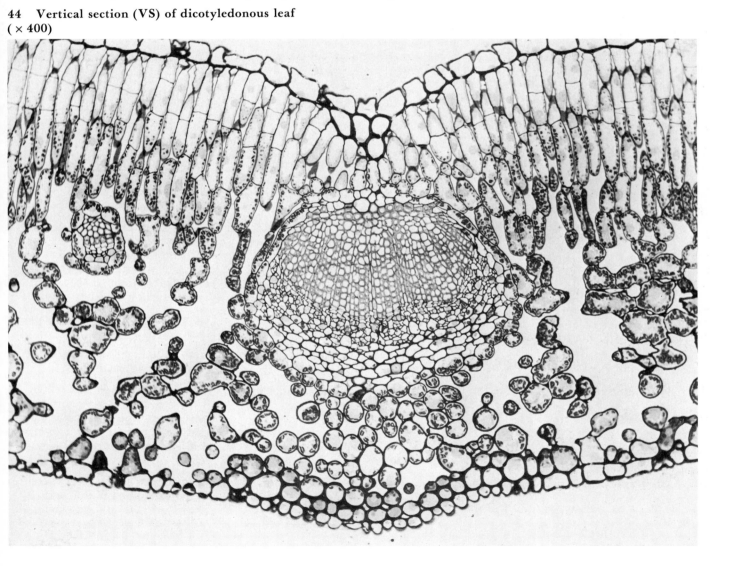

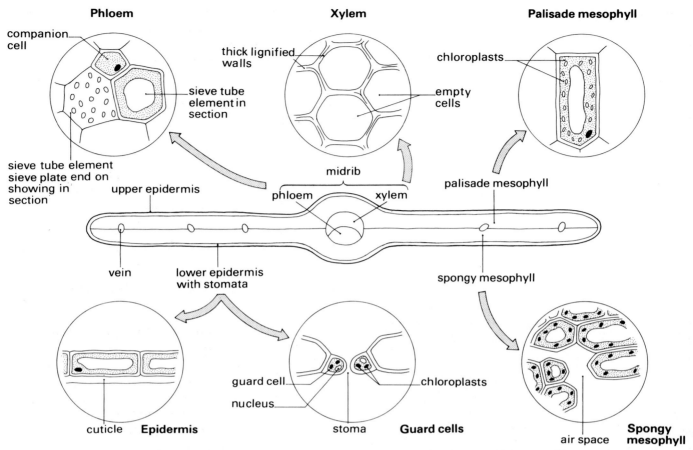

**Phloem**

companion cell

thick lignified walls

sieve tube element in section

sieve tube element sieve plate end on showing in section

upper epidermis

**Xylem**

empty cells

midrib

phloem    xylem

**Palisade mesophyll**

chloroplasts

palisade mesophyll

vein

lower epidermis with stomata

spongy mesophyll

cuticle    **Epidermis**

guard cell

nucleus

stoma    **Guard cells**

chloroplasts

air space    **Spongy mesophyll**

**45    Tissue map of a typical leaf with diagrams showing the structure of component cells as seen under high power**

**46    Structure and function of leaves**

| Cell type | Function |
| --- | --- |
| Epidermal layer | Protects against mechanical injury and entry of pathogens. The waxy cuticle it secretes reduces water loss. Transparent to allow light to reach mesophyll. |
| Stomata | Control of gas exchange and water loss. |
| Palisade mesophyll | Cells contain chloroplasts which carry out photosynthesis. |
| Spongy mesophyll | Cells large and well spaced out. They provide a system through which substances in solution and gases can pass. Fewer chloroplasts but some photosynthesis occurs here. |
| Vascular bundle | Transport of water and minerals in in the xylem and products of photosynthesis in the phloem. |

leaves. All leaves have a common pattern of cell types included in the following regions.

(*a*) Epidermis with guard cells (stomata)

(*b*) Palisade mesophyll

(*c*) Spongy mesophyll

(*d*) Vascular bundles with xylem and phloem

## 2.7.3 Structure and function of leaves

The function of cells types found in the leaf is listed in figure 46.

### 2.7.4 Sun and shade leaves

The following photographs show sections through the blade of a leaf from the upper part of *Fagus* (beech tree) (figure 47) and the lower shaded part of the tree (figure 48).

*SAQ 47* What is the main structural difference between the two photographs?

*SAQ 48* What would be the advantage to the shade plant of this difference?

*SAQ 49* Of the two leaves, which will have the lower compensation point? What is the advantage to the leaf concerned?

**47 Transverse section (TS) of *Fagus* sun leaf (× 400)**

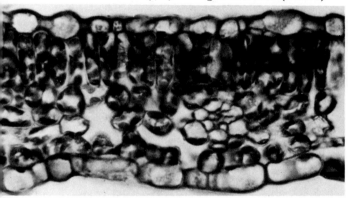

**48 TS of *Fagus* shade leaf (× 400)**

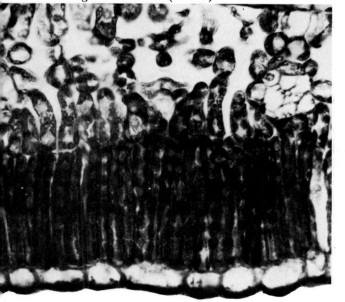

### Practical H: The internal structure of leaves – transverse sections

#### Materials

Microscope and lamp, prepared slides – transverse sections of leaves

#### Procedure

Examine each prepared slide of a transverse section of a leaf, using first low and then high power.

(*a*) Make a low power plan to show the arrangement of tissues in a leaf. (N.B. No individual cells should be drawn.)

(*b*) Make high power drawings of 2–3 representative cells from each region of the leaf.

In your observations and drawings, pay especial attention to the following.
(i)   Cell size and shape
(ii)  Wall thickness
(iii) Nature of cell contents
(iv)  Presence or absence of chloroplasts
(v)   Presence of intercellular spaces

Label your drawings fully and annotate to explain the functions of the cell types.

Show this work to your tutor.

### 2.7.5 Chloroplast structure and function

The basic structure and function of chloroplasts was outlined in the *Cells and the origin of life* unit. Figures 49 and 50 summarise the relation between structure and function.

**49   Two-dimensional view of the structure of a
chloroplast**

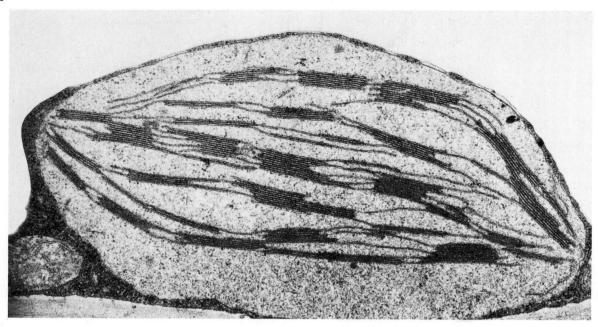

**50   Chloroplast structure and function**

lamellae of granal and
intergranal regions bear
photosynthetic pigments
and present a large
surface area for the
absorption of light

double membrane
separates chloroplast
from cytoplasm;
controls internal
movement of $CO_2$,
water and minerals and
outward movement of
sugars etc.

stroma – fluid-filled space
containing enzymes of
light-independent stage of
photosynthesis; in close
contact with lamellae where
ATP and $NADPH_2$ (products
of the light-dependent stage
required by the light-independent
stage) are produced

intergranal and granal
lamellae bear
cytochromes and other
electron carriers
associated with the
light-dependent
reactions; lamellae act
as convenient surfaces
for spacing of molecules
involved in electron flow

## 2.7.6 Summary assignment for section 2.7

**1** Using figure 45 as a basis, make labelled drawings
to show leaf structure. Include a low power tissue
map and high power drawings to show individual cell
types.

**2** Answer questions (a)–(l) as a plan for the following
examination question.

Explain how a leaf appears to be suited to carry out
photosynthesis. (London, January 1969)

What advantages or disadvantages are there in
having
(a) a thin leaf?
(b) stomata?
(c) more stomata on the lower surface?
(d) photosynthetic cells below rather than on the
surface?

(e) photosynthetic cells only just below the surface?
(f) an epidermal layer?
(g) an epidermal layer which lets light through?
(h) a waxy cuticle?
(i) xylem and phloem cells below photosynthetic cells?
(j) flattened cells in the epidermal layer?
(k) long photosynthetic cells at right-angles to epidermal layer?
(l) an opening and closing mechanism in the stomata?

3 Using figure 50 as a basis, make a summary diagram to outline the relationship between structure and function in chloroplasts.

Show this work to your tutor.

Self test 7, page 90, covers section 2.7 of this unit.

## 2.8 The role of minerals in plant nutrition

The carbon, hydrogen and oxygen of the organic molecules in plants are obtained in the form of carbon dioxide and water during the process of photosynthesis.

Other elements are required by plants in the form of dissolved mineral salts obtained from soil water. This section is concerned with a study of these elements.

### 2.8.1 Mineral requirements

At least 13 elements have been shown to be required for plant growth and development. These are known as **essential elements**. Some of these elements are required in very small quantities and are consequently known as **trace elements**.

The technique which has been developed to reveal which elements are essential to a plant is the water culture technique. This involves growing a plant in a solution made up of distilled water and the purest salts available. The effect of leaving out or adding particular elements can then be observed.

One of the major reasons for the investment in research into the mineral nutrition of plants has been a need to improve the productivity of crops to supply the increasing world population. The results obtained apply largely to crop plants which are flowering plants, and care must be taken in applying these results to other plants.

Table 51 lists the essential elements which have been discovered so far and summarises the effects of deficiency in these elements and their supposed role in a plant. Study the table carefully and answer the following question.

**SAQ 50** What elements are required for the following?
(a) Protein structure
(b) Chlorophyll structure and formation
(c) Middle lamella
(d) Cell membranes
(e) Enzyme structure or activation
(f) Reactions involving ATP

---

### Practical I: Demonstrating the importance of certain elements in plant growth

---

The water culture technique described in section 2.8.1 can be used to demonstrate the effects of mineral deficiency on plant growth.

#### Materials

40 ml complete plant culture medium, 40 ml each of plant culture media lacking in one of the following minerals: phosphorus, nitrogen, calcium, potassium, magnesium, iron, sulphur, 8 test-tubes, test-tube rack, cotton wool, aluminium foil, 8 identical seedlings of beans, peas, wheat or maize

#### Procedure

(a) Seedlings which have recently germinated are placed in a series of water culture tubes as shown in figure 52.

(b) The aluminium foil excludes light from the culture solution preventing the growth of algae.

(c) The tubes are placed so that each one receives an equal amount of light.

(d) The seedlings are left for 3–4 weeks, the culture

| Element | Role in the plant | Effect of deficiency |
|---|---|---|
| Nitrogen ($NO_3^-$) | Absorbed as nitrate. An essential constituent of amino acids, proteins, nucleotides, chlorophyll, coenzymes (e.g. NAD and NADP). | Chlorosis (yellowing of leaves). Small-sized plants. |
| Phosphorus ($H_2PO_4^-$) | Absorbed as orthophosphate ($H_2PO_4^-$). Essential constituent of ATP and nucleic acids, and as phospholipids in cell membranes | Small-sized plants. Leaves a dull, dark green. Development of necrotic (brown) areas in leaves and petioles. |
| Sulphur ($SO_4^{2-}$) | Absorbed as sulphate. Involved in formation of sulphur bridges in protein structure. | Chlorosis. (**High** levels of sulphur in polluted regions may lead to reduced growth.) |
| Potassium ($K^+$) | Required for protein synthesis. Helps maintain general ionic balance. Activates many enzymes (e.g. pyruvate kinase in glycolysis). | Weak, spindly plants. Undersized seeds. |
| Calcium ($Ca^{2+}$) | Formation of middle lamella (between cell walls). | Stunted growth, especially of roots. |
| Magnesium ($Mg^{2+}$) | Acts as a cofactor for most enzymes using or making ATP. Essential part of chlorophyll structure. Helps maintain general ionic balance. | Chlorosis. Reduced growth. Short internodes. Inhibition of flowering. |
| Iron ($Fe^{2+}$ or $Fe^{3+}$) | Oxidation/reduction function as active centre constituent of electron carriers in photosynthesis and respiration and in nitrogen-fixing enzymes. Involved in chlorophyll formation. | Chlorosis and stunted growth. |
| Boron ($BO_3^{3-}$) | Required for efficient translocation of sugars. Influences calcium uptake and use. | Translocation blocked. Disorganisation of meristems. Flowering suppressed. Fruits and seeds abnormal. |

*Trace elements (involved in action and/or structure of enzymes)*

| Element | Role in the plant | Effect of deficiency |
|---|---|---|
| Manganese ($Mn^{2+}$) | Activates some enzymes (hydrolases and carboxylases). Constituent of oxygen-producing unit in photosynthesis. | Inability to form certain proteins. Chlorosis. Necrosis. |
| Copper ($Cu^{2+}$ or $Cu^+$) | Essential constituents in active centre of several redox enzymes and proteins (phenol oxidases, cytochrome oxidase). | Chlorosis. Die-back of shoots. |
| Zinc ($Zn^{2+}$) | Constituent of several enymes (e.g. carbonic anhydrase, alcohol dehydrogenases). | Inability to form carbonic anhydrase. Failure of leaves to expand and stems to elongate. IAA (indole acetic acid, a growth hormone) levels low. |
| Molybdenum ($MoO_4$) | Constituent of nitrogen-fixing enzymes. | Inability to use nitrate. |
| Chlorine ($Cl^-$) | Involved in oxygen-producing unit in photosynthesis. Helps maintain general ionic balance. | Reduced rate of photosynthesis. |

solution being topped-up with the appropriate mix. Their growth should be monitored at regular intervals, such as weekly.

## Results and discussion

1 Make any observations and measurements you feel relevant to the effects of mineral deficiencies on plant

**52　A water culture tube**

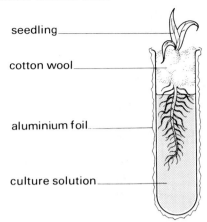

seedling

cotton wool

aluminium foil

culture solution

growth. Record these results.

**2** What precautions did you need to take when carrying out this experiment?

**3** Why is it necessary to prevent the growth of algae in the culture solutions?

**4** Discuss the shortcomings of this experimental technique.

Show this work to your tutor.

### 2.8.2 Summary assignment for section 2.8

**1** Explain the following terms:
essential element, trace element, water culture.

**2** List the eight essential plant elements. For each, state its role in the plant and its major deficiency syndrome.

**3** Name five trace elements and state their general function.

Show this work to your tutor.

Self test 8, page 91, covers section 2.8 of this unit.

Write an essay plan to answer the following past examination question. Then write the essay.

Describe how you would investigate the mineral requirements of a green plant. List these requirements and explain the physiological importance of any four of the required ions. (London, January 1975)

Hand the plan and the essay to your tutor.

### 2.8.3 References for section 2.8

*Plants at Work* by F. C. Steward, chapter 4.
*The Life of the Green Plant* by A. W. Galston, chapter 3.
*Plants and Mineral Salts* by J. F. Sutcliffe & D. A. Baker, Studies in Biology No. 48.

# Section 3  Heterotrophic nutrition

## 3.1 Introduction and objectives

In this section you will study the ways in which heterotrophic organisms obtain their food. You will learn what the basic requirements of their diet are and examine the ways in which the food is processed by the organism.

At the end of this section you should be able to do the following.

(*a*) Name three feeding categories into which animals may be grouped.

(*b*) State the sources of particulate food for animals.

(*c*) Give a brief description of the feeding method of *Amoeba*.

(*d*) Explain concisely what is meant by filter feeding, giving a short summary of filter feeding in a bivalve mollusc and *Daphnia*.

(*e*) Name three other animals from different classes, other than those in (*d*) above, that filter feed.

(*f*) Describe briefly how *Arenicola* obtains food.

(*g*) List five other methods of obtaining food, naming the special structures involved in each case.

(*h*) Name five fluid feeders and their sources of food.

(*i*) Give an illustrated account of the structure of mammalian teeth.

(*j*) List the ways in which the teeth and jaws of herbivorous and carnivorous mammals are adapted to their diets.

(*k*) Give an account of the method by which a locust obtains food.

(*l*) List the main organic and inorganic nutrients and account for their importance in the metabolism of heterotrophs.

(*m*) State the factors which affect the energy requirements of an individual.

(*n*) Show what is meant by a balanced diet by assessing the nutritional value of a meal.

(*o*) Define the following terms: ingestion, digestion, absorption, assimilation, egestion.

(*p*) Identify the main regions of the gut of a mammal.

(*q*) Label a diagram showing a section through the gut of a mammal.

(*r*) Explain the meaning of mechanical digestion and describe some examples of it.

(*s*) Describe the process of chemical digestion, to include details of the main enzymes involved and the end-products of this digestion.

(*t*) Explain how the small intestine is suited to its function of absorption.

(*u*) Outline the structure and importance of the liver.

(*v*) List the products of digestion which are absorbed into the body and give examples of what happens to them.

## 3.2 Feeding methods

---

### AV 4: Methods of feeding

---

In this section you will be introduced to a range of feeding methods used by different animals.

**Materials**

VCR and monitor

ABAL video sequence: *Methods of feeding*
Worksheets

## Procedure

(*a*) Check that you have all the relevant materials for this activity.

(*b*) Check that the video cassette is set up ready to show the appropriate sequence – *Methods of feeding*.

(*c*) Start the VCR and stop it to complete the work-sheets as instructed.

(*d*) If you do not understand anything, stop the video, rewind and study the relevant material again before consulting with your tutor.

(*e*) If possible, work through the video and work-sheets with a small group and discuss the material with your fellow students.

Show this work to your tutor.

---

## Practical J: Circus of experiments on feeding in invertebrates

---

The following practical investigations will enable you to find out more about feeding in snails, locusts and flies. For each organism studied you should make annotated diagrams of your findings in your practical notebooks.

### To investigate feeding in snails

### Materials

Pulmonate pond snails in a glass tank whose walls are encrusted with algae, magnifying lens × 10, prepared slide of snail radula, microscope and lamp

### Procedure

(*a*) Observe algal encrusted walls of the tank for feeding trails (see figure 53). Use a magnifying lens to observe the snails as they feed.

(*b*) Examine the prepared slide of the radula.

### 53 Limpet feeding trails

Describe the feeding apparatus of the snails by means of annotated diagrams.

### To investigate feeding in locusts

### Materials

Locusts, living and preserved, grass or grated carrot, video: *Problems of feeding – the locust as a biting insect*, forceps, mounted needles, small scissors, microscope (binocular or with × 4 objective) or lens and lamp, cork mat or small dissecting dish, pins

### Procedure

(*a*) Watch living locusts feeding on grass or other plant material. Notice how quickly they eat through the food. Make notes on your observations. In particular, try to discover answers to the following questions.
(i) What sense organs are present?
(ii) What role do the legs play in feeding?
(iii) Does the whole body move during feeding?
(iv) What is the function of the palps?
(v) Where does food enter?
(vi) How do the mandibles move?
(vii) What is the function of the labrum?

(*b*) Watch the video which shows *Schistocerca gregaria* (the desert locust).

(*c*) Pin a preserved locust on its back on a cork mat. Identify the mouthparts as shown in figure 54.

(*d*) Try to answer the following questions during your subsequent observations.

## 54  Side view of a locust head showing mouthparts

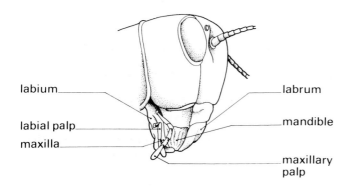

(i) Why are the mandibles so hard?
(ii) Is there any evidence of the muscles which cause the mandibles to move?
(iii) Do the cutting edges of the mandibles overlap when they are closed?
(iv) Is there any evidence of wear on the cutting edges?

## 55  Inner surface of left mandible

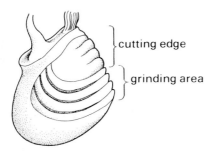

(e) Fold back the labrum or upper lip. Tap the large black jaws or mandibles with a mounted needle. The maxillae which function as accessory jaws lie just below the mandibles. The labium, a shovel-like lower lip, forms the floor of the mouth.

(f) Remove the mouthparts one by one, noting their relationship to each other. Examine with a binocular microscope or the × 4 objective of your microscope.

(g) Draw the mouthparts and annotate your diagrams to explain how each is involved in feeding.

## To investigate feeding in the fly

### Materials

Slide of mouthparts of fly, microscope and lamp, video: *Problems of feeding – the housefly*

### Procedure

(a) During your observations in (b) and (c), look for answers to the following questions:
(i) What shape is the feeding organ?
(ii) Where is the mouth?
(iii) What happens to the feeding organ when the fly is not feeding?
(iv) Suggest a function for the maxillary palps.
(b) View the video *Problems of feeding – the housefly*.

(c) Observe the prepared slide showing the mouthparts of a fly.

(d) Make diagrams to illustrate your observations. Annotate your diagrams to explain the feeding mechanism.

Flies feed on a wide variety of foods. They are incapable of eating solid material. Therefore, digestion begins outside the body when saliva is pumped onto the food to bring about its liquefaction.

When the fly lands on food and the proboscis is extended, by the combined action of muscles and blood pressure, saliva flows out from the fleshy lobes onto the food substrate, moistening and partly dissolving it. Powerful pharyngeal muscles cause the semi-liquid food to be sucked up the oesophagus by their pumping action.

The proboscis of the fly consists of a single hollow tube, hinged where it joins the head, thus it can be tucked away when not in use. The oesophagus passes down the proboscis, and ducts from the salivary gland open into it on each side. The proboscis ends in two fleshy lobes which are penetrated by food canals.

**56** Photomicrograph of the mouthparts of a blowfly

## 3.2.1 Feeding in mammals

Mammals are characterised by the possession of teeth set in the upper and lower jaws. The teeth and moveable lower jaw provide a useful tool for catching and chewing food, or both.

The structure of a mammalian tooth is shown in the figure 57.

Mammalian teeth are different from the teeth of other vertebrates since they include several types, each specialised for a particular function. Such a dentition is described as **heterodont**.

The four main tooth types are shown in the diagrams in figure 58.

A mammal has two sets of teeth during its life. The first set (milk teeth) consist of incisors, canines and premolars. The second, or permanent, dentition also includes molars.

A record of the number of teeth of each type in the upper and lower jaw on one side may be presented by the **dental formula**. Teeth are represented by their initial letter written in lower case. Those teeth in the upper jaw are written above the line, those in the lower jaw below the line. The dental formula for an adult human is

$$i.\frac{2}{2} \quad c.\frac{1}{1} \quad pm.\frac{2}{2} \quad m.\frac{3}{3}$$

Use figure 57 to answer the following questions.

*SAQ 51* Which substance(s) give teeth their strength?

*SAQ 52* Which parts of a tooth contain living cells?

*SAQ 53* Which structures hold a tooth in its bony socket?

*SAQ 54* What is the function of (*a*) the gums, (*b*) the blood vessels associated with the teeth?

*SAQ 55* When having a filling, which region of the tooth has the dentist's drill reached when you feel some discomfort? Why do you feel this discomfort?

*SAQ 56* Which characteristics from the list below apply to each of the four teeth types?
(*a*) incisors (*b*) canines (*c*) premolars (*d*) molars

Crown (i) with two cusps
     (ii) flat with sharp edge
     (iii) with four to five cusps
     (iv) with one prominent cusp

Roots (v) two or three
     (vi) one or two
     (vii) one

*SAQ 57* Write the dental formula for a child with a full milk dentition. How many teeth will such a child possess?

---

**Practical K: Circus of experiments on feeding in mammals**

---

The following practical investigations will enable you to find out more about feeding in sheep, cats and humans. For each organism studied, you should make annotated diagrams of your findings in your practical notebooks.

**To investigate the teeth, skull and lower jaw structure of a herbivore – the sheep**

**Materials**

Video: *Problems of feeding – chewing in herbivores*, sheep skull and lower jaw

## 57 Structure of a mammalian tooth

**enamel:** hardest material of the body, 96% mineral matter arranged in cylinders aligned perpendicularly to surface

**pulp cavity:** contains the nerve and blood supply to the tooth

**cement:** a form of bone which surrounds the neck and root of the tooth

**nerves and blood vessels:** nerves carry sensory information to the brain; blood vessels supply oxygen and nutrients and remove waste products of metabolism

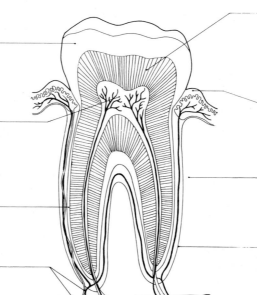

**dentine:** bulk of tooth organic fibrillar network 70% mineral matter, mainly calcium; harder than bone (ivory); perforated by canals in which lie the cells that produce the dentine

**gum or gingiva:** protects jawbone and roots of teeth

**alveolar bone:** provides a socket for the roots of the tooth

**periodontal membrane:** strong fibrous connective tissue which connects cement-covered root to bone of socket or alveolus

## 58 Tooth types of mammals

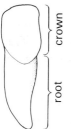

crown

root

**incisor:** at front of mouth; in upper jaw carried by a separate bone, the maxilla; edges flat and sharp

**canine:** crown has a prominent point or cusp; roots large and deeply embedded (name originates from well-developed form in the dog – *Canis*)

**premolar:** crown has two cusps; one or two roots

**molar:** these crowns have four or five cusps; upper molars have three roots, lower ones have two

all teeth shown from front view

## Procedure

(*a*) View the video once. Observe the following.

(i) How the sheep picks up pellets

(ii) Type of jaw movement shown by various herbivores (for instance, is the movement directly up and down or from side to side).

(*b*) Examine skull and lower jaw (mandible). The dental formula of a sheep is

$$\text{i.}\frac{0}{3} \quad \text{c.}\frac{0}{1} \quad \text{pm.}\frac{3}{3} \quad \text{m.}\frac{3}{3} = 32$$

(i) Identify these teeth on your specimen.

(ii) The premaxilla is covered by a horny pad in the living animal. Bearing this in mind, how can the sheep crop grass? Which teeth will be concerned with this function?

(iii) Note the gap between the canines and premolars. This is the diastema. This allows the tongue to manipulate food and keep freshly cropped grass separate from that being chewed.

(iv) Suggest a reason why the molars and premolars are very similar in structure.

(v) Measure the distance between the outer edges of

$m_2$ on the left and right sides of the mandible and compare it with the distances between the outer edges of $m_2$ in the upper jaw. Relate this to the type of jaw movement you observed when watching the video.

(vi) Align the upper and lower teeth on one side of the upper and lower jaw. Note down the way the cheek teeth (pm and m) fit together.

## 9  Skull of a sheep

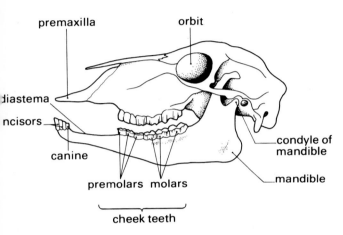

(vii) Examine the surfaces of the cheek teeth. Compare them with the diagram of a generalised mammalian tooth (figure 57). Identify the different materials of which the teeth are composed and use this information to explain how the ridged pattern you observe has arisen. Feel the teeth and assess how effective they are likely to be at chewing grass.

Grass contains silica and can be very tough material to chew. The cheek teeth are prevented from becoming entirely worn away because they grow throughout the life of the animal. This is possible because the roots remain open, admitting a continual supply of food and oxygen, via the blood vessels, to the pulp cavity.

(viii) Look at the teeth sideways and see if you can find the guiding ridges which enable the teeth to slot together effectively during chewing movement. Note that these ridges are not the same as the main cutting ridges which stand out on the upper surface.

60  (a) Molar tooth of a young herbivore – longitudinal section (LS)
(b) Molar tooth of a herbivore showing the ridges developed by grinding – LS

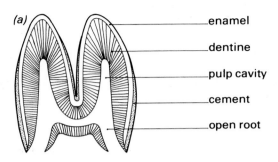

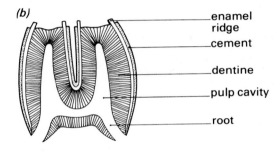

(ix) Look at the point of articulation of the mandible with the skull. This joint is known as a **condyle**. The articulating surfaces are flat and allow the lower joint to rotate in a horizontal plane.

(x) Make an annotated drawing of the skull and mandible of a sheep, incorporating the points you have observed so far.

(xi) View the video again and add any additional information you think may be relevant.

(xii) Make annotated sketches of the ridged, surface of a cheek tooth.

## To investigate the teeth, skull and lower jaw structure of a carnivore – the cat

### Materials

Video: *The problem of feeding – chewing in carnivores*, cat skull or skull of larger member of cat family

## Procedure

(a) View the video several times.

(i) How does a cat pick up food? Which teeth are used for this?

(ii) Watch carefully as the cat chews its food. How does it do this?

(iii) What is the role of the tongue?

(iv) How does it deal with larger pieces of food?

(v) Note down the differences in feeding that you can observe in the lion and tiger.

(vi) Watch the movements of the cat skull and try to relate this to your observations on feeding in the cat.

(b) Examine the skull and mandible.

The dental formula of the cat is

$$\text{i.}\frac{3}{3} \quad \text{c.}\frac{1}{1} \quad \text{pm.}\frac{3}{2} \quad \text{m.}\frac{1}{1} = 30$$

(a) Identify these teeth on your specimen. What is unusual about $pm^1$ and $m^1$? Which type of tooth is particularly well developed? Feel them with the soft pad of your finger and compare the sensation of this and the other teeth. Suggest the role of these teeth.

(b) Assess the relative size of the incisors compared with the other teeth. What role do they play in feeding?

## 61  Cat skull and mandible

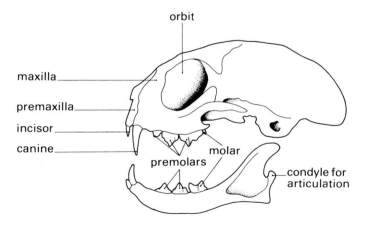

(c) Look very carefully at $pm^3$ and $m_1$. Slot the mandible into its condyle and see how these teeth function. If necessary, review the last section of the

video which shows the relationship of these teeth. These teeth are know as **carnassial** (flesh-cutting) **teeth**. Note that the inner side of $pm^3$ and the outer side of $m_1$ are flat and that they articulate together like the blades of scissors.

It is these teeth which are brought into action when the animal works its food to the back of its mouth and chews on each side alternately.

(d) Measure the distance between $pm^3$ on each side of the skull and $m_1$ on each side of the mandible. Use this to explain why the cat shifts its food from side to side.

(e) Is there any evidence of grinding teeth? Give reasons for your answer.

(f) Look at the condyles between the mandible and skull. It forms a hinge-joint which enables the jaw to open and close but does not allow much freedom for other movements. Why is a joint of this kind useful in carnivores?

(g) Notice the large size and forward-facing position of the orbits. Relate this to the mode of feeding of the animal.

(h) Make an annotated drawing of the skull and mandible of a cat incorporating the points you have observed so far.

(i) Again, view the video and add any additional information you consider relevant.

(j) Make annotated sketches to show the mode of functioning of the carnassial teeth.

(k) Compare your work with that of other students in your class. Be prepared to offer and receive constructive criticisms of the work.

**To investigate the teeth, skull and lower jaw structure of an omnivore – human**

### Materials

Skull of man (may be plastic), samples of teeth, mirror

## Procedure

(*a*) Study the specimens provided, together with your own teeth which you can examine with the mirror. Try to find answers to the following questions.
(i) To what extent do the canines differ from the incisors?
(ii) What functions would you expect the two types of teeth to have?
(iii) Describe the cheek teeth.
(iv) How are the cheek teeth adapted for an omnivorous diet?
(v) Examine the joint between the mandible and the skull. Compare it with the joint in the sheep and the cat.

(*b*) Make annotated drawings of your observations showing how the human teeth and jaws are adapted to an omnivorous diet.

Show this work to your tutor.

---

**62   Human skull from side and front**

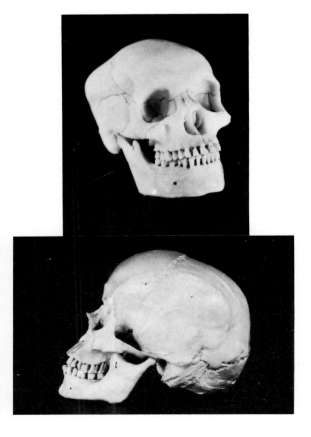

## 3.2.2 Summary assignment for section 3.2

Write an essay plan to answer the following past examination question.

Discuss the range of feeding methods in animals. (London, January 1980)

Find a quiet place and set yourself 35 min to write the essay.

Show this work to your tutor.

Self test 9, page 91, covers section 3.2 of this unit.

## 3.3 Food

This section is concerned with an examination of the essential requirements for a healthy diet.

### 3.3.1 Basic requirements of the diet

The diet must provide the body with energy, body-building materials and all the necessary basic materials for a healthy life with the exception of oxygen. **Carbohydrates** and **fats** are the main sources of energy and their role in respiration will be explored in the next section. A limited amount of carbohydrate is stored as glycogen and the excess is converted into fat which can be stored in unlimited quantities, mainly under the skin and mesenteries. Cereals and potatoes are a good source of carbo-hydrates. Fats are found in foods such as cheese, eggs and nuts.

**Protein** is the main body-building component of the diet. It cannot be stored in the body and any excess is converted into urea and carbohydrate. Meat, fish, milk products and beans are good sources of protein.

**Vitamins** and **minerals** are required in small quantities and yet have a very important role to play. Some have a structural role, such as calcium in bone, while others, such as sodium, are necessary for maintaining a correct osmotic balance in the body.

**Fibre** or **roughage** has a number of important func-tions during the digestive and absorptive phases of nutrition. The final requirement is **water**, which may be taken in as fluid or as a component of food.

Water is also produced as a result of the biochemical reactions of respiration. It is essential (a) as a structural component of cells and hence of the body, (b) as a medium for transport within the body and (c) as a medium for biochemical reactions.

### 3.3.2 A balanced diet

**A balanced diet** is one which provides adequate amounts of proteins and all the necessary vitamins and minerals as well as meeting energy requirements. Most foods contain several nutrients and most minerals and vitamins are present in many foods that are in common use. Thus, eating a varied diet containing a wide selection of foods, will usually meet nutritional requirements.

### 3.3.3 Food and energy

In heterotrophs, food provides the source of energy enabling the organism to carry out its metabolic activities. The energy value of a food source is measured in terms of kilojoules (kJ). Previously, the units kilocalories were used (popularly written as the Calorie with a capital letter). Some books still refer to this term. One kilocalorie is equivalent to 4.184 kJ.

The rate at which energy is used up while the body is lying down, still, relaxed and warm is known as the **basal metabolic rate** (BMR). The BMR for a young man of average build and temperament, weighing 70 kg is about 7000 kJ per day. The average young woman of 55 kg has a BMR of about 6000 kJ. The BMR for both sexes tends to fall with increasing age.

All individuals engage in some kind of activity which will bring their energy requirements above that for the BMR. For instance, sitting requires 63 kJ h$^{-1}$, washing dishes 247 kJ h$^{-1}$, house painting 606–609 kJ h$^{-1}$ and sawing wood 1756 kJ h$^{-1}$.

Daily energy requirements of a variety of people are expressed visually in figure 63.

**63    Daily energy requirements of various people**

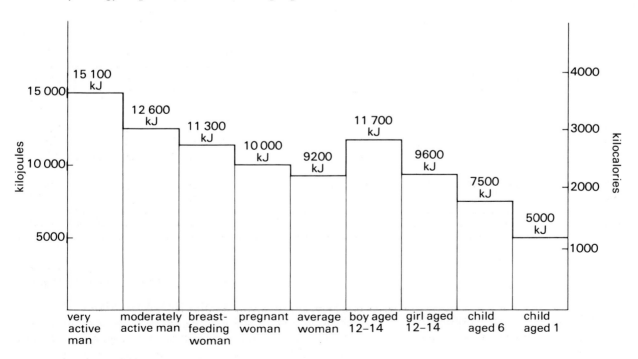

| | Energy | Protein Recommended | Protein Minimum requirement | Calcium | Iron | Vitamin A (retinol equiv.) | Thiamin | Ribo-flavin | Nicotinic acid equiv. | Vitamin C | Vitamin D |
|---|---|---|---|---|---|---|---|---|---|---|---|
| | (kJ) | (g) | (g) | (mg) | (mg) | (mg) | (mg) | (mg) | (mg) | (mg) | (µg) |
| Girl 15–17 | 9600 | 58 | 40 | 600 | 15 | 750 | 0.9 | 1.4 | 16 | 30 | 2.5 |
| Boy 15–17 | 12 500 | 75 | 50 | 600 | 15 | 750 | 1.2 | 1.7 | 19 | 30 | 2.5 |

65   Comparison of the nutritional value of a snack and a cooked meal

| | Mass | Energy | Protein | Fat | Carbo-hydrate | Calcium | Iron | Vitamin A (retinol equiv.) | Thiamin | Ribo-flavin | Nico-tinic acid equiv. | Vitamin C | Vitamin D |
|---|---|---|---|---|---|---|---|---|---|---|---|---|---|
| | (g) | (kJ) | (g) | (g) | (g) | (mg) | (mg) | (µg) | (mg) | (mg) | (mg) | (mg) | (µg) |
| **Snack meal** | | | | | | | | | | | | | |
| Bread, white | 113 | 1205 | 9.2 | 2.0 | 61.6 | 112 | 2.0 | 0 | 0.20 | 0.04 | 2.8 | 0 | 0 |
| Butter | 14 | 435 | 0.1 | 11.5 | 0 | 2 | 0 | 141 | 0 | 0 | 0 | 0 | 0.18 |
| Cheese | 57 | 935 | 14.4 | 19.6 | 0 | 460 | 0.4 | 238 | 0.02 | 0.28 | 3.0 | 0 | 0.20 |
| Lettuce | 14 | 4 | 0.2 | 0 | 0.2 | 3 | 0.2 | 24 | 0.01 | 0.01 | 0.1 | 2 | 0 |
| Tomato | 28 | 12.5 | 0.2 | 0 | 0.7 | 4 | 0.1 | 33 | 0.02 | 0.01 | 0.2 | 6 | 0 |
| Coffee, instant | 3 | 17 | 0.1 | 0 | 1.0 | 4 | 0.1 | 0 | 0 | 0 | 1.3 | 0 | 0 |
| Milk (summer) | 57 | 159 | 1.8 | 2.2 | 2.8 | 68 | 0 | 24 | 0.02 | 0.08 | 0.6 | 1 | 0.02 |
| TOTAL | 286 | 2767.5 | 26.0 | 35.3 | 66.3 | 653 | 2.8 | 460 | 0.27 | 0.42 | 8.0 | 9 | 0.40 |
| **Cooked meal** | | | | | | | | | | | | | |
| Lamb, roast | 71 | 870 | 16.2 | 15.8 | 0 | 5 | 1.5 | 0 | 0.05 | 0.18 | 6.5 | 0 | 0 |
| Peas, boiled | 57 | 109 | 2.8 | 0 | 4.4 | 8 | 0.6 | 28 | 0.14 | 0.06 | 1.4 | 8 | 0 |
| Chips, fried | 85 | 854 | 3.3 | 7.8 | 31.8 | 12 | 1.2 | 0 | 0.09 | 0.03 | 1.8 | 6–18 | 0 |
| Peaches, canned | 113 | 418 | 0.4 | 0 | 26.0 | 4 | 2.0 | 48 | 0 | 0.04 | 0.8 | 4 | 0 |
| Custard | 85 | 326 | 2.7 | 3 | 11.1 | 93 | 0 | 30 | 0.06 | 0.12 | 0.6 | 0 | 0.03 |
| TOTAL | 411 | 2577 | 25.4 | 26.6 | 73.3 | 122 | 5.3 | 106 | 0.34 | 0.43 | 11.1 | 18–30 | 0.03 |

*SAQ 58* Using the information given in this section, state three important factors which affect the daily energy requirements of an individual.

### 3.3.4 Analysis of the nutritional content of diets

Figure 64 shows the recommended daily intake of nutrients for a girl and boy aged 15–17 years. Figure 65 shows the nutritional value of two possible choices for a midday meal. Study these figures carefully.

**SAQ 59** Which of the two meals do you consider would have the highest nutritional value for a girl or boy (state which) aged between 15 and 17 years? Give your reasons.

**SAQ 60** Which of the nutritional requirements are inadequate for each meal?

**SAQ 61** Suggest any inexpensive additions to each meal which would improve its nutritional value. Use figure 66 as a guide to help you.

**66 Sources of energy and nutrients**

| | |
|---|---|
| Energy | Sugar, lard, margarine, butter, white bread, old potatoes, brown bread, biscuits, breakfast cereals |
| Protein | White bread, brown bread, milk, cheese, old potatoes, baked beans, chicken, eggs, breakfast cereals, liver, fish |
| Carbohydrate | Sugar, white bread, old potatoes, brown bread, breakfast cereals, new potatoes, biscuits |
| Calcium | Milk, cheese, white bread, ice cream, brown bread, carrots |
| Iron | Liver, canned beans, old potatoes, brown bread, white bread, breakfast cereals, frozen peas, fresh green vegetables |
| Vitamin A | Carrots, liver, margarine, butter, cheese, milk |
| Thiamin | Fortified breakfast cereals, potatoes, bread, frozen peas, milk, pork |
| Riboflavin | Liver, fortified breakfast cereals, milk, eggs, ice cream, cheese, potatoes |
| Nicotinic acid | Fortified breakfast cereals, potatoes, liver, white bread, chicken, brown bread, milk, frozen peas |
| Vitamin C | Fruit juices, oranges, new potatoes, fresh green vegetables, tomatoes, frozen peas |
| Vitamin D | Margarine, fatty fish, butter, eggs |

### 3.3.5 Carbohydrates, lipids and proteins

Figure 67 is a revision of carbohydrates, lipids and proteins which acts as a pre-test. Much of the work you will have covered in the *Cells and the origin of life* unit.

Read through the chart, copy it out and fill in the gaps. The answers are on page 110, at the end of the answers to the SAQs.

If you get less than 45 marks, before going on to the next section you should revise the section in *Cells and the origin of life* or read an alternative account in a standard textbook, such as *Biology* by John W. Kimball.

### 3.3.6 Inorganic nutrients

The body contains about 20 inorganic elements, called **minerals**, which must be derived from food. Eight of these should be present in relatively large amounts. These are nitrogen, calcium, phosphorus, potassium, sulphur, sodium, chlorine and magnesium. The remainder, which are essential for normal metabolism, are needed in much smaller quantities, and are known as **trace elements**. These include iron, cobalt, copper, zinc, vanadium, chromium, molybdenum, manganese, silicon, tin, selenium, fluorine and iodine. Minerals are required for four main purposes:

(*a*) as constituents of bones and teeth (for instance calcium, phosphorus and magnesium);
(*b*) as constituents of body cells (iron, phosphorus, sulphur and potassium);
(*c*) as soluble salts which give the body fluids their composition and stability, which are essential for life (sodium, potassium, chlorine);
(*d*) as factors involved in chemical reactions in the body, including those concerned with the release of energy during metabolism (iron, phosphorus and magnesium).

In addition, some minerals have specific purposes in the body. Iodine is a constituent of the hormone thyroxin which plays a part in regulating the metabolic rate of the body; cobalt is a constituent of vitamine $B_{12}$. Copper and zinc are present in various

| Food | Chemical composition | Types | Functions |
|---|---|---|---|
| Carbohydrates | Contain the elements (1) _____ (2) _____ and (3) _____ . <br><br> The basic unit is a simple sugar. <br><br> In more complex carbohydrates, the bond between sugar molecules is called a (4) _____ bond. <br> Reducing sugars, e.g. (5) _____ , _____ , _____ , differ from non-reducing sugars, e.g. (6) _____ ,in that only the former have a free (7) _____ group. | (a) Monosaccharides, e.g. galactose, (21) _____ and (22) _____ . <br> (b) Disaccharides, e.g. lactose, (23) _____ and (24) _____ . <br> (c) Polysaccharides, e.g. (25) _____ ,(26) _____ , (27) _____ , chitin and pectin. | (a) Immediate energy supply, e.g. (35) _____ . <br><br> (b) Energy store (36) _____ and (37) _____ . <br> (c) Structured components (i) of plant cell walls (38) _____ . (ii) of insect cuticles (39) _____ . |
| Lipids | Contain the elements (8) _____ (9) _____ and (10) _____ . <br><br> The basic unit is a (11) _____ . This is composed of (12) _____ and (13) _____ . | (a) Simple lipids (triglycerides) e.g. (28) _____ ,(29) _____ and (30) _____ . <br> (b) Complex lipids, e.g. (31) _____ and (32) _____ . <br> (c) Steroids. | (a) Energy source (higher energy content per unit weight than carbohydate). <br> (b) Protection (i) against mechanical damage e.g. (40) _____ ; (ii) against heat, e.g. (41) _____ ; (iii) against water loss, e.g. (42) _____ . <br> (c) Structural, e.g. cell membrane. |
| Proteins | Contain the elements (14) _____ , (15) _____ , (16) _____ , (17) _____ , and (18) _____ . <br> The basic unit is an (19) _____ . <br> The basic units are linked together by (20) _____ bonds. | (a) Fibrous proteins, e.g. keratin (skin and hair), collagen (bone), actin and myosin (muscle), fibrin (blood clotting). <br> (b) Globular protein, e.g. enzymes, haemoglobin, antibodies, hormones, e.g. insulin. <br> Of the 20 amino acids required for proteins, eight are required ready made in the diet of adults (nine in growing children). These are the (33) _____ . <br> The remainder, the (34) _____ _____ , can be synthesised from other amino acids by mammals. | (a) Structural functions, e.g. (43) _____ , (44) _____ and (45) _____ . <br> (b) Movement functions, e.g. (46) _____ . <br> (c) Functions associated with maintaining metabolism, e.g. (47) _____ (48) _____ (49) _____ ,and (50) _____ . |

| Vitamin | Source | Function | Effect of deficiency |
| --- | --- | --- | --- |
| A<br>Retinol | Fish liver oils, liver, kidney, dairy produce and eggs. Vegetable products (e.g. spinach, carrots) contain carotene and other carotenoids. These may be converted to vitamin A during absorption in the small intestine. | Essential for vision in dim light. Maintenance of healthy skin and surface tissues, especially those which secrete mucus. | Night blindness. Severe eye lesions (xerophthalmia). Increased chance of respiratory and intestinal infections. Roughening of the skin. |
| $B_1$<br>Thiamin | Wholemeal bread and whole grain cereals, yeast extract. Offal. | Steady and continuous release of energy from carbohydrate. (Component of coenzyme co-carboxylase.) | Beriberi (polyneuritis). Weakness of limbs, abnormal sensations, possible paralysis. Cardiovascular system may be affected. Oedema in legs. |
| $B_2$<br>Riboflavin | Milk, meat especially liver, eggs. | Essential for utilisation of energy from food (produces FP – a hydrogen acceptor important in cellular respiration). | Rare in man but includes sores at corner of mouth, lips may ulcerate, tongue purplish-red. May be dermatitis. |
| Nicotinic acid (Niacin) | Instant coffee, liver, meat, fish, tea, eggs, legumes. | Also involved in utilisation of food energy. Essential part of NAD and NADP (hydrogen acceptors). | Pellagra, in which skin becomes dark and scaly, especially where it is exposed to light. Diarrhoea. Possibly confusion and hysteria. |
| $B_6$<br>Pyroxidine | Meat, fish, eggs, whole cereals. | Involved in metabolism of amino acids. Also for the formation of haemoglobin. | Excessive production of urea. |
| $B_{12}$<br>Cyanoco-balamin | Only in animal products. Liver, eggs, cheese, meat. | A mixture of several compounds, all of which contain cobalt. With folic acid it is needed by rapidly dividing cells such as those in bone marrow which form blood. | Pernicious anaemia and the degeneration of nerve cells. |
| Folic acid | Offal and raw green leafy vegetables, pulses, bread, oranges, bananas. | Action with $B_{12}$ in rapidly dividing cells and other functions. | Megaloblastic anaemia – particularly during pregnancy. (Mild deficiency is widespread in Britain.) |
| C<br>Ascorbic acid | Rose-hips, blackcurrants, citrus fruits. Raw fruits and vegetables. | Maintenance of healthy connective tissue. Role in biological redox reactions. | Bleeding from small blood vessels and into the gums. Wounds heal more slowly. Scurvy leads to death. |
| D<br>Calciferols | Cod liver oil, fish, margarine. Action of sunlight on skin. | Necessary for maintaining levels of calcium and phosphorus in blood – enhances absorption of dietary calcium from intestine and regulates interchange of calcium between blood and bone. | Rickets – deformed limb bones too weak to support weight of body. |
| E | Vegetable oils, cereals, etc. | Not known for man – concerned with fertility in rats. | Anaemia? Destruction of liver cells |

nzymes and zinc is associated with the hormone
insulin; while iron is necessary for the formation of
haemoglobin.

By consulting the references given below, answer the
following question.

Describe the importance to mammals of the following
minerals. Three or four concise sentences will be
sufficient for each substance.

(i) iron                    (iv) magnesium
(ii) phosphate              (v) potassium
(iii) calcium               (vi) chloride

Name four other minerals required by animals.

*References*

*Success in Nutrition* by Magnus Pike.
*Plant and Animal Biology*, vol. 2, by A. E. Vines &
N. Rees.
*Animal Physiology* by Knut Schmidt-Nielsen.

## 3.3.7 Vitamins

Until the end of the nineteenth century, it was
thought that a diet was adequate if sufficient protein,
fat, carbohydrate and inorganic elements were
supplied. This view changed when it was shown that
natural unrefined foods contained substances
essential for life and health which the body is unable
to synthesise for itself. These organic substances were
called vitamins and were found to be present in very
minute quantities in foods. Vitamins A, D, E and K
are found mainly in fatty food and are called fat-
soluble vitamins, while the vitamins of the B group
and vitamin C are soluble in water.

Although vitamins do not provide energy or body-
building materials, they are essential for energy
transformations and the regulation of metabolism.

Some form parts of enzyme systems and operate
within narrow limits of temperature and pH. Figure
68 provides details of the source and function of
vitamins.

Use figure 68 to answer the following questions.

*SAQ 62* What additions to the diet would you
recommend to an individual suffering from anaemia?

State which vitamins might be deficient.

*SAQ 63* Blindness, as a result of nutritional
deficiency, occurs in Third World countries. What
could cure this condition.?

*SAQ 64* Which vitamins are involved in energy
reactions within the cell?

*SAQ 65* What are the functions of the fat-soluble
vitamin which can be formed by the action of
sunlight on the skin?

## 3.3.8 Fibre in the diet

Fibre, originally known as roughage, is an important
component of the diet. It mainly comprises the
cellulose of plant cell walls. Therefore, it is plentiful
in plant products, especially cereals.

It has little direct nutritional value, but is important
because it is indigestible and absorbs water in the intestine. This increases the volume of the faeces and
helps prevent constipation by encouraging its movement through the colon and rectum.

---

**Practical L: Investigating the chemical
constituents of foods**

---

To determine the nutritional value of foods it is
necessary to perform tests to detect the presence of
the different chemical constituents. In the following
practicals you can try out such tests and then use
them to assess the composition of foods.

In the first practical you are given solutions or
suspensions of the five major food types, together
with instructions for carrying out tests to identify
these types.

### Materials

$10 \text{ cm}^3$ 1% starch solution, $10 \text{ cm}^3$ 10% glucose
solution, $10 \text{ cm}^3$ 10% sucrose solution, $10 \text{ cm}^3$ 1%
albumen solution, $10 \text{ cm}^3$ vegetable oil, iodine
solution, ☠ Benedict's solution, ☠ 2M hydrochloric
acid, solid sodium hydrogen carbonate, ☠ 10M
sodium hydroxide solution, ☠ 10% copper sulphate
solution, ethanol, test-tube rack, 5 test-tubes, Bunsen
burner, test-tube holder, spatula, dropping pipette,
labels, beaker of water to act as a water-bath

## Procedure

(*a*) Take a sample of each food type and try out all the tests on it as described in figure 69.

(*b*) Make a written report of your results and show it to your tutor.

### Extension practical on food tests

You may use the above tests to investigate the chemical constitution of various foods. For solid substances you will need to make a suspension of the food in water before doing the test.

Make a full written report of your investigations and show it to your tutor.

You can read about the background chemistry to these tests in *Practical Plant Physiology* by J. Roberts & D. G. Whitehouse.

### 3.3.9 Summary assignment for section 3.3

**1** Explain the following terms:
basal metabolic rate, kilojoule, balanced diet, vitamin.

**2** Outline the major functions of the following: carbohydrates, proteins, lipids, vitamins A, B, C and D, fibre, water.

**3** Construct a chart based on figure 69 which summarises the tests used to identify chemical

### 69   Food identification tests

| Food | Reagent | Test | Positive result |
|------|---------|------|-----------------|
| starch | iodine soln. | Add 5 drops iodine soln. to 1 cm³ of test soln. | Dark blue-black colour. |
| reducing sugar | Benedict's soln. (caustic) | Add 1 cm³ Benedict's soln. to 2 cm³ test soln. Place in a water-bath and heat gently, shaking all the time. Remove from heat when mixture boils. | Red-orange colour. Brown or green colour indicates decreasing amounts of sugar. |
| non-reducing sugar | HCl Benedict's soln. | Add 3 drops dilute HCl to 2 cm³ test soln. Boil for 2 min. Cool. Neutralise with conc. NaOH. Add 1 cm³ Benedict's soln. Heat gently over a low flame, shaking all the time. Remove from the heat when mixture boils. | Red-orange colour. Brown or green colour indicates decreasing amounts of sugar. N.B. When testing an unknown substance, always do the reducing sugar test first. It is only possible to detect a non-reducing sugar in the presence of a reducing sugar by comparing the colour of the two tests. |
| protein (Biuret test) | sodium hydroxide soln. copper sulphate soln. | Add 2 cm³ sodium hydroxide to 1cm³ test soln. Add 1 drop copper sulphate soln. down the side of the test-tube. | A blue ring at the surface indicates a protein. On shaking, the ring disappears and the solution turns purple. |
| lipids | isopropanol alcohol | Add 3–4 drops of test soln. to 2 cm³ isopropanol alcohol and shake to dissolve. Filter if necessary. Add to 2 cm³ water. | Cloudy white emulsion. |

omponents in foods, together with a description of positive results.

Show this work to your tutor.

Self test 10, page 92, covers section 3.3 of this unit.

## 3.4 Processing food

The food that is taken in by living organisms must first be broken down physically and chemically into a form which can be absorbed from the gut into the blood. It can then be transported by the blood to those sites in the body where it is required. This section deals with these processes.

### 3.4.1 The gut

Animals get food into the mouth in a variety of different ways, as you have seen. The mouth is the entrance to a long tube known as the gut, alimentary canal or digestive tract. It is along this tube that the food then passes.

In the less advanced multicellular animals, such as *Hydra* and flatworms, there is only one opening or mouth which serves both for taking in food and expelling waste. The more advanced animals also have a posterior opening, or anus, for the exit of waste.

As the food passes along the gut it must be broken down or digested into small soluble molecules. This may occur in two ways.

**Mechanical digestion** is the breakdown of bulky solid food into small pieces by the chewing action of teeth and the churning action of the stomach wall muscles.

**Chemical digestion** is the breakdown of large molecules into smaller ones by hydrolysis catalysed by specific digestive enzymes.

The small soluble molecules are then absorbed into the bloodstream. Different regions of the gut are specialised to perform the various functions of digestion and absorption. The incorporation and use of the absorbed products of digestion by the body is known as **assimilation**.

Figure 70 summarises the four main stages of nutrition in relation to the gut.

### 3.4.2 The human gut

The diagrammatic gut in figure 70 is a simple, straight tube of uniform diameter. In many animals, including humans, the gut is coiled, thus allowing a greater length to be accommodated within the body. Also, it often varies in thickness and other details along its length associated with the functions of the different regions.

Figure 71 shows the human gut with its associated glands.

*SAQ 66* List the main regions of the gut from mouth to anus. Ignore associated glands and side-branches.

70  Nutrition and the gut

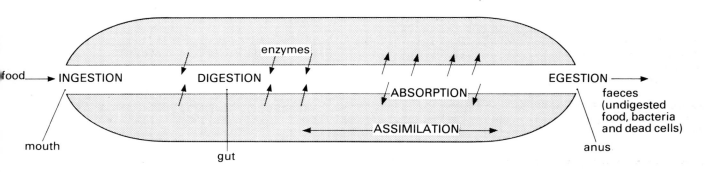

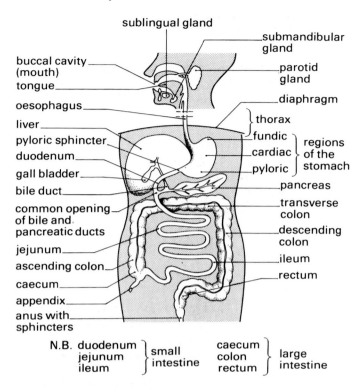

buccal cavity (mouth)
tongue
oesophagus
liver
pyloric sphincter
duodenum
gall bladder
bile duct
common opening of bile and pancreatic ducts
jejunum
ascending colon
caecum
appendix
anus with sphincters

sublingual gland
submandibular gland
parotid gland
diaphragm
thorax
fundic
cardiac } regions of the stomach
pyloric
pancreas
transverse colon
descending colon
ileum
rectum

N.B. duodenum
jejunum  } small intestine
ileum

caecum
colon  } large intestine
rectum

*SAQ 67* Which part of the gut is represented by a small, blind-ended side-branch?

*SAQ 68* Into which region of the gut do the bile duct and pancreatic duct open?

*SAQ 69* Which region of the gut (a) has the widest diameter, (b) has sacculated walls?

A transverse section through the gut wall shows that the wall is made up of four main layers: mucosa, submucosa, external muscle coat and serosa.

The composition of these layers together with their functions is shown in figure 72.

*SAQ 70* Which layers of the gut wall are concerned with
(a) support,
(b) transport of digested food away from the gut?

*SAQ 71* What would you expect to happen to the diameter of the gut tube if
(a) the longitudinal muscles relaxed and the circular muscles contracted?

(b) the longitudinal muscles contracted and the circular muscles relaxed?

### 3.4.3 Movement of food through the gut

Food is moved along the gut by a process known as **peristalsis**. Each portion of swallowed food, a bolus, stretches the gut, and the circular muscles behind it contract and push the bolus onward into a region where the muscles are relaxed (see figure 73).

Successive waves of contraction and relaxation move the food along the gut. Relaxation of the circular muscles is brought about by the longitudinal muscles. The muscle layers are **antagonistic**, one layer contracting and bringing about the relaxation of the other layer. Peristaltic movements occur throughout the gut and may be particularly noticeable in the stomach which, when food is present, is constantly changing its shape. Food is churned and mixed in this large, expansible storage 'bag'. Large peristaltic waves pass over the stomach at the rate of three every minute, and are especially strong in the pyloric region, which ends in the sphincter muscle. This muscle opens at intervals to release food to the first part of the small intestine, the duodenum.

In addition to peristaltic movements of the gut, there are contractions of the circular muscles at intervals along the length of the small intestine known as **segmentation** movements. These serve to mix the gut contents by pushing food backwards and forwards. The pattern of segmentation in the small intestine alters about 20 times a minute and gives the intestine the appearance of a string of sausages.

### Practical M: Dissection and display of the alimentary canal of a mouse

### Materials

Dissection dish, dissecting instruments: scalpel, forceps, scissors, pins, *Dissection Guide III, The Rat* by H. G. Q. Rowett, mouse (preferably freshly killed), Ringer's solution at about 37 °C in a water-bath

**2  Diagram showing the main regions of the gut wall**

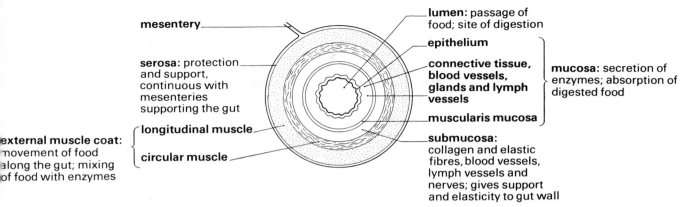

mesentery

serosa: protection and support, continuous with mesenteries supporting the gut

external muscle coat: movement of food along the gut; mixing of food with enzymes

longitudinal muscle

circular muscle

lumen: passage of food; site of digestion

epithelium

connective tissue, blood vessels, glands and lymph vessels

muscularis mucosa

mucosa: secretion of enzymes; absorption of digested food

submucosa: collagen and elastic fibres, blood vessels, lymph vessels and nerves; gives support and elasticity to gut wall

**3  Peristalsis – moving food along the gut**

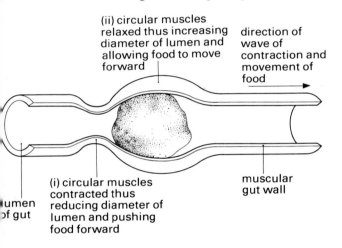

(ii) circular muscles relaxed thus increasing diameter of lumen and allowing food to move forward

direction of wave of contraction and movement of food

(i) circular muscles contracted thus reducing diameter of lumen and pushing food forward

muscular gut wall

lumen of gut

**Procedure**

Follow the instructions laid down in the dissection guide. The mouse is generally dissected under water to wash away blood from capillary bleeding. Ordinary small pins should be used to pin the mouse to the dish. When pins are used to display organs, they should be placed **against** and not through the organ.

If a freshly killed mouse is used follow parts (a)–(f), otherwise start at (g).

(a) As soon as the mammal has been killed, place it on its back in a dissecting dish and quickly fix in position with pins. Immediately cover the animal with warm Ringer's solution taken from the water-bath.

(b) Open the abdomen of the mouse, working as swiftly as possible, using the instructions in the dissection guide.

(c) As soon as the flaps of skin and muscle have been pinned back, stop dissecting and observe the contents of the abdomen very carefully. Use a binocular microscope and look for movement of any kind.

Check temperature and replace Ringer's solution, if necessary. If you see movements, record your observations very carefully.

(d) Try to identify the main parts of the alimentary canal. If your exploration triggers more movements, stop and observe these.

(There are at least three types of distinct patterns of movement known to occur in the gut.)

(e) Answer the following questions in your practical notebook.

1 Which parts of the gut appeared to move? How do you explain these movements?

2 Do gut movements appear to move food from mouth to anus? If not, what appeared to be the effect of the movements on the food in the gut?

(f) Continue with your dissection.

(g) Make a carefully labelled drawing of the mouse when you have opened the abdominal cavity, with the intestines in situ, but with any masking fat removed. Annotate this drawing with any observations that may be relevant. Make sure you have identified everything you have drawn.

(h) When the intestines have been displayed, make another carefully labelled drawing of your dissection. Identify any broken or damaged parts when labelling.

Show this work to your tutor.

---

### 3.4.4 Early investigations into digestion

In 1882, a Canadian fur-trapper was accidentally shot. He recovered but a hole remained in his side, leading to the interior of his stomach. He agreed to cooperate with a US army surgeon in a series of experiments to investigate the process of digestion.

The following is an account of one of the experiments performed.

'11th January. At 3 o'clock p.m. dined on bread and eight ounces of recently salted lean beef, four ounces of potatoes, and four ounces of turnips, boiled. In fifteen minutes, took out a portion of the contents of the stomach. The meat made its appearance in an incipient stage of digestion.

At 3 o'clock, 45 minutes, took out another portion. The meat and bread only appeared in a still more advanced stage of digestion.

The texture of the meat was, at this time, broken into small shreds, soft and pulpy, and the fluid containing it had become more opaque and quite gruel-like, or rather glutinous in appearance.

I put this second parcel in a vial and placed it in water, on the sand bath, at the temperature of the stomach, (100° Fahrenheit), as indicated by the thermometer immediately preceding its extraction, and continued it there.

At 5 o'clock, took out another quantity. Digestion had advanced in about the same ratio as from the first to the second time of extracting; and when compared with the second parcel, contained in the vial on the bath, little or no difference could be perceived in them; both were nearly in the same stage of digestion. That contained in the vial had advanced regularly and rapidly; nearly all the particles of meat had disappeared, become chymified, and changed into a reddish-brown sediment, suspended in the more fluid parts, with small particles, resembling loose, white coagulae, floating about near the surface.

On taking out the third parcel, small pieces of vegetables appeared in a partial stage of digestion. This was also put into a vial and placed on a bath, with the second parcel, and the same uniform temperature (100°) kept up with frequent, gentle agitation.

At 6 o'clock p.m., digestion had progressed equally in both. The only difference to be seen was the particles of vegetables in a less-advanced stage than the meat.

The contents of both vials, kept on the bath, and nearly in the same temperature, until the next morning, were completely digested, except the few small particles of vegetables which remained almost entire.

The contents of the vials at this time were of the consistence of thin jelly, and of a lightish brown colour; tasting peculiarly insipid, saltish and acid. After standing at rest awhile, the brownish sediment subsided towards the bottom, while small particles of whitish coloured, loose coagulae floated about in the fluid above. The undigested particles of vegetables settled to the bottom.'

*SAQ 72* (a) Which food was the first to be changed in the stomach?
(b) Describe the stage of digestion of the salted lean beef.
(c) Give the evidence for the assertion that the gastric juice appears to be the main agent of digestion in this region.
(d) Suggest a reason for the vials being placed in a sand bath at 100 °F.
(e) Which of the foods ingested appears the most difficult for the stomach to digest. Suggest a reason for this.
(f) The contents of the stomach after a period of digestion are called chyme. Describe this chyme.

### 3.4.5 Chemical digestion

We now know that digestion is brought about by a

series of enzymes. These enzymes are secreted by glands. The glands may either empty into the lumen of the gut through ducts, for instance the salivary glands and pancreas, or they may lie in the wall of the gut itself, such as the gastric glands and the intestinal glands in the walls of the stomach and intestine respectively.

Enzymes are sensitive to pH. Acid or alkaline secretions are liberated into certain regions of the gut where they create the correct conditions for enzyme action.

Digestive enzymes are not produced continuously. Their flow is stimulated by the presence of food in the gut. An outline of this control is shown in figure 74.

### 74 Control of secretion of digestive juices

| Digestive juice | Stimulus for secretion |
| --- | --- |
| Salivary juices | Presence of food in mouth, sight, smell or thought of food |
| Gastric juices | Mechanical or chemical stimulation of stomach wall |
| HCl in stomach | The hormone gastrin produced by presence of food in stomach |
| Intestinal juices | Presence of food in small intestine |
| Pancreatic juice and bile | Several hormones (secretin, pancreomysin and cholecystokinin) produced by presence of acidified food in intestine |

As food passes along the gut, it mixes with the enzymes and is broken down. Proteins are hydrolysed by stages into polypeptides, dipeptides and finally amino acids; fats are hydrolysed into fatty acids and glycerol; starches are hydrolysed into disaccharides and monosaccharides.

A summary of the digestive secretions and their functions is shown in figure 75.

*SAQ 73* In which region of the gut does digestion of the following food types begin?
(*a*) fat    (*b*) disaccharide

(*c*) starch    (*d*) protein

*SAQ 74* In which region of the gut would you find the following?
(*a*) hydrochloric acid
(*b*) amylase (1,4-glucan-4-glucanohydrolase)
(*c*) rennin (chymosin)
(*d*) peptidase (peptide amino-acid hydrolase)

*SAQ 75* Several digestive enzymes are produced in inactive form and must be activated before they are effective in digestion. Name two such enzymes and state the reactions they catalyse.

*SAQ 76* (*a*) Name the substance important in fat digestion which does not contain any enzymes.
(*b*) Where is this substance produced?

## Practical N: Investigating enzyme activity in the gut

The presence of any enzyme is detected by its activity, the ability to catalyse a certain chemical change. This practical looks for the presence of certain enzymes in gut extracts by their ability to break down their specific substrates. An amylase will hydrolyse starch, a protease protein, a lipase fats and cellulase cellulose. If the substrates are incorporated into agar, enzymes can diffuse through the agar to hydrolyse the substrate. This can be detected by the use of reagents which stain the original substrates.

### Materials

BDH universal indicator, 4 petri dishes containing (i) agar with Marvel milk powder as substrate (ii) agar with starch/iodine substrate (iii) agar with mayonnaise (iv) agar with blended filter paper substrate, freshly-killed *Lumbricus*, locust, rat or mouse, cork-borer 5 or 6, dissecting dish and instruments (scissors, scalpel and forceps), incubator, 10% copper sulphate solution, compasses or dividers, iodine solution

### Procedure

(*a*) Dissect out the gut of the animal you are

| Secretion | Where effective | Where produced | Enzymes present | Enzyme substrate | Enzyme product | Notes |
|---|---|---|---|---|---|---|
| saliva | mouth (and temporarily in stomach) | salivary glands | amylase (1,4-glucan-4-glucanohydrolase) | cooked starch | maltose | Water to soften food. Mucus to make food slippery. Salts to provide neutral medium for action of amylase and to help preserve teeth from acids formed by bacteria. |
| gastric juice | stomach | gastric glands | pepsin (peptide peptidohydrolase) | protein | polypeptides | Water to soften food. Mucus to prevent enzymes damaging stomach lining. HCl to stop amylase working, to allow pepsin to work and kill germs. |
|  |  |  | rennin (chymosin) (not secreted by adult mammals) | milk protein | curdled milk |  |
|  |  |  | lipase | fat | fatty acids and glycerol | Pepsin is first secreted in the inactive form, pepsinogen. HCl converts it to pepsin. |
| bile | small intestine (duodenum) | liver (stored in gall bladder) |  |  |  | Water to soften food. Bile pigments – waste excreted in faeces. Bile salts – alkaline; they neutralise acidity of chyme, stopping the action of pepsin, but promoting the action of intestinal enzymes. Emulsifies fats. |
| pancreatic juice | pancreas | small intestine (duodenum) | lipase | fat | fatty acids and glycerol | Water. Alkaline salts help increase the alkalinity in the intestine and combine with fatty acids to form 'soaps'. |
|  |  |  | amylase (1,4-glucan-4-glucanohydrolase) | starch | maltose |  |
|  |  |  | trypsin (peptide peptido-hydrolase) | polypeptides | dipeptides | Trypsin is first secreted in the inactive form, trypsinogen. |
| intestinal juices (*succus entericus*) | intestinal glands | small intestine | enterokinase (enteropeptidase) | trypsinogen | trypsin | Water. Mucus to protect intestinal wall. |
|  |  |  | peptidase (peptide aminoacido-hydrolase) | polypeptides and dipeptides | amino acids |  |
|  |  |  | maltase (α-0-glycosidase) | maltose | glucose |  |
|  |  |  | sucrase (sucrose α-0-glucohydrolase) | sucrose | glucose and fructose |  |
|  |  |  | lactase (β-0-galactosidase) | lactose | glucose and galactose |  |
|  |  |  | nucleotidases | nucleotides | pentose sugars, phosphoric acid and organic bases |  |

investigating.

Instructions for dissecting an earthworm or locust are given below. You may also refer to Rowett's dissection guides.

*Earthworm dissection*

(i) Find the darker dorsal surface. The anterior end is pointed while the tail end is somewhat flattened.

(ii) Hold the worm in your left hand and with fine, sharp scissors make a small longitudinal cut in the body wall just above the clitellum (see figure 76). Take care to cut as shallowly as possible.

**76   How to open the worm**

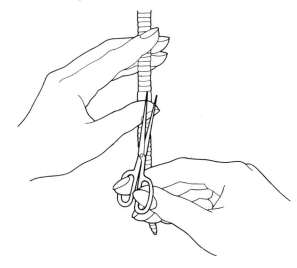

(iii) Hold the scissors so that the points slope upwards. Cut forward to the prostomium.

(iv) Place a pin through the pharynx into the dissecting dish and pull gently backwards to stretch the tissues.

(v) Pin again through the tail so that the worm is fully extended.

(vi) Hold the body wall open by paired pins.

(vii) Use forceps to lift the gut in the region of the gizzard and carefully cut the septa so that the gut can be freed from underlying tissues. Work forward gently, as the intestine is easily torn.

(viii) When you come to the oesophageal region,

place a finger on the lobes of the seminal vesicles (see figure 77) and carefully pull the oesophagus from the median seminal vesicle.

(ix) Cut across the pharynx, posterior to the nerve collar to release the gut.

(x) Yellow material (chloragogenous cells) may be removed from the surface of the intestine with a brush.

**77   Dissection of earthworm gut**

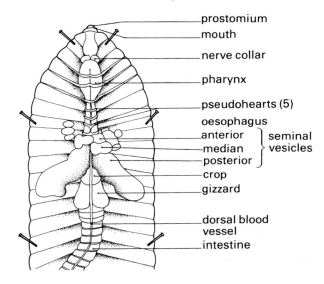

*Locust dissection*

(i) Remove the wings and legs.

(ii) Place the locust in the dissection dish, ventral side downwards.

(iii) Pin through the posterior end of abdomen.

(iv) Hold the neck in position by angled pins on either side.

(v) Using fine-pointed scissors, make a cut in the membrane between two segments of the abdomen. Make two parallel cuts as shown in figure 78. Hold the scissors at an angle so that the points slope upwards while cutting. This prevents damage to the underlying tissues.

(vi) Remove the dorsal strip of the exoskeleton,

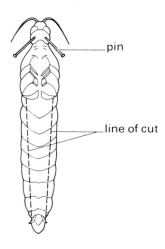

separating it from underlying tissues with a fine scalpel.

(vii) Remove the pearly grey air sacs, yellow fat and muscles where necessary.

**79    Dissection of locust gut**

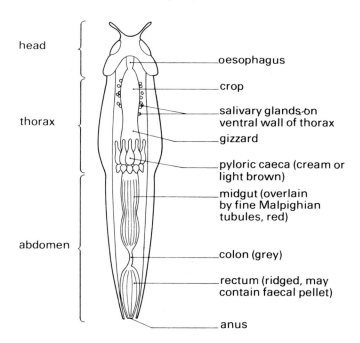

head

thorax

abdomen

oesophagus

crop

salivary glands on ventral wall of thorax

gizzard

pyloric caeca (cream or light brown)

midgut (overlain by fine Malpighian tubules, red)

colon (grey)

rectum (ridged, may contain faecal pellet)

anus

(viii) Cut through the gut just before it enters the head and through the end of the rectum. Carefully remove gut from the animal.

(b) Cut six wells in each agar plate using the cork-borer, tilting the borer slightly when withdrawing the unwanted agar.

(c) Label the wells with small, numbered paper labels (see figure 80).

(d) Transfer small parts of the gut to the numbered wells (5 mm length of gut of an invertebrate or 5 mm² piece of gut wall of a vertebrate).

(e) Add distilled water to each well up to the level of the gel surface.

(f) Incubate the agar and Marvel plates at 25 °C for about 12 h. The plates with cellulose (filter paper) and mayonnaise are best incubated at 30 °C: cellulose for 12 h, mayonnaise for 24 h. Satisfactory results can be obtained from the starch–iodine plate in 6 h.

**80    Apparatus set up for investigation of enzyme activity in the gut**

(g) If sufficient gut material is available, a rough pH measurement for each part of the digestive tract can be obtained by placing a segment of gut wall into BDH universal indicator for about 10 min and then comparing its colour with that of the standards. (This will not indicate pH values below 4.0.)

(h) (i) When incubation time is complete, enzyme activity can be assessed in the starch and Marvel plates by measuring the area of the clear zones around the wells (graph paper, compasses or dividers may be useful for this).
(ii) Lipase activity in the mayonnaise plate can be more clearly shown by flooding the plate with 10% copper sulphate solution which should be left for 30 min to penetrate the gel. Zones containing fatty acids stain blue-green after this treatment and stand

out clearly against the white, opaque background.
(iii) Areas of cellulase activity can be seen by flooding the plates with a strong solution of iodine in KI (2 g iodine: 4 g KI in 100 cm$^3$ water) which should be left to penetrate the gel for about 30 min and then washed off under a running tap. Zones without cellulose become pale brown or colourless while other areas remain dark brown or purple.

(i) Draw up a table of results showing the area of the gel cleared by enzymes in mm$^2$ for each region of gut. Present this information in suitable graphical form also. Remember, amylase (1,4-glucan-4-glucanohydrolase) digests starch, protease digests Marvel, a protein substance, lipase digests the fat present in mayonnaise and cellulose in filter paper is digested by cellulase. If you have results for pH, these should be included in the table.

(j) Write a discussion of the results of this investigation, bearing in mind the following questions.

**1** Which area of gut is most actively concerned with digestion?

**2** From your results, does it appear that amylases are produced in one region only, or more than one? Suggest the main areas producing amylase in your specimen.

**3** Is there any evidence that pH is affecting the functioning of any enzyme in any region?

**4** Which area of the gut of the species you investigated is most likely to be concerned with absorption? Justify your answer.

**5** Does the amount of cellulase present suggest that plant material is likely to be an important source of food for the animal. Is its production limited to any region?

**6** Does any pattern emerge for the distribution of lipase?

Show this work to your tutor.

## 3.4.6 Absorption of digested food

By the time food reaches the ileum, digestion is complete and the main phase of absorption begins.

There is some absorption in other parts of the gut, for instance alcohol and some drugs (such as Disprin) are absorbed in the stomach and water and certain vitamins are taken up in the large intestine.

The ileum is well adapted to its function of absorption in three main ways.
(a) Its surface area is very large.
(b) The cells lining the walls contain large numbers of mitochondria which make energy available for the uptake of digested food.
(c) The wall contains an extensive transport system of blood and lymph capillaries for picking up the digested food and carrying it to different parts of the body where required.

## Practical O: To investigate the structure of the ileum wall

In this practical you will examine the structure of the ileum wall and consider how this structure is related to its functions.

### Materials

Microscope and lamp, prepared slide of TS through the ileum of a mammal

### Procedure

(a) Examine the slide under lower power and high power and identify the main regions of the ileum wall.
(b) Make a LP tissue map to show the main regions of the wall.
(c) Examine the epithelium layer and make drawings to show the structure of
(i) a columnar epithelial cell,
(ii) a goblet cell,
(iii) an enzyme-producing cell.

Indicate where these cells are found on your LP map.

Refer to figures 72 and 82 to help you interpret what you see.

(*d*) Calculate the width of a TS of the ileum. Using the method given in the Introductory unit, estimate the length of the surface of a villus and the increase of surface provided by the villi.

(*e*) On your diagrams indicate
(i) glands secreting digestive enzymes,
(ii) cells which protect the ileum wall against self-digestion,
(iii) those features of the wall associated with increasing its surface area,
(iv) those features of the wall associated with transport of digested food.

Show this work to your tutor.

---

Figure 81 is a photograph of a section through the ileum wall. The inner surface can be seen to be covered by finger-like projections known as villi. The thin, dark interconnected threads visible within the wall and villi are blood vessels. Figure 82 is a diagram interpreting the main structures of the ileum wall.

Study these figures carefully.

*SAQ 77* Which structures are responsible for increasing the surface area of the ileum wall for the absorption of digested food products?

*SAQ 78* What is the name of the glands which secrete the intestinal juice?

*SAQ 79* Name the structures in the villi which you would expect to be responsible for carrying away the absorbed products of digestion.

### 81   Section through ileum wall

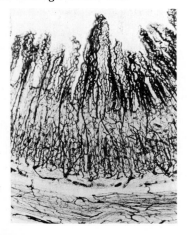

## 82   Microscopic structure of the wall of the ileum

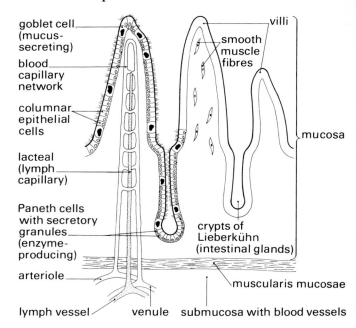

The smooth muscle fibres in the villi enable these structures to move constantly. This brings them into contact with a greater number of food molecules and hence increases absorption.

Figure 83 is an electron micrograph of two epithelial cells from the ileum lining. Examine the plasma membrane in the region bordering the lumen of the ileum (upper region in photograph).

*SAQ 80* What is unusual about the plasma membrane in this region and how will this affect absorption of sugars and other foods?

*SAQ 81* Name the organelles present in large numbers in the cell. How might they be involved in the absorption of sugars and other foods?

The exact mechanism of absorption of the products of digestion is not known. It is not just a case of diffusion. It is at least in part a selective process since those sugars used in metabolism are absorbed more rapidly than those which are not metabolically important. There is evidence of several transport systems involving carrier molecules and requiring energy.

You can read more about digestion and absorption in

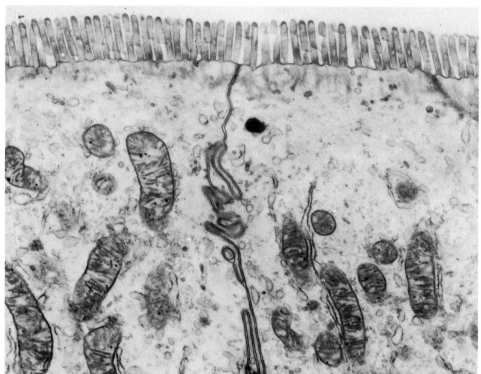

*The Lining of the Small Intestine* by Florence Moog (*Scientific American*, November 1981).

Use the QS3R system of note-taking to help you read this article.

### 3.4.7 Transport of absorbed products of digestion

Monosaccharides and amino acids together with some products of fat digestion are absorbed into blood capillaries. Much of the fatty acids and glycerol are absorbed into epithelial cells, re-synthesised into fats and then passed into the lacteals as finely suspended droplets, giving a milky appearance.

The blood capillaries carry digested food to the hepatic portal vein which leads directly to the liver.

The lacteals lead to lymph vessels which eventually empty their contents into the bloodstream in the neck region.

### 3.4.8 Absorption of water and egestion

The material passing into the large intestine from the ileum contains undigested matter, mostly of vegetable origin (cellulose) in man. This roughage plays an important role in the healthy functioning of the large intestine as it provides bulk which aids peristaltic movements in the large intestine. However, water is the major constituent and the food remains are very fluid. Also present are bacteria, mucus, dead cells sloughed off from the lining of the alimentary canal and bile pigments. The first parts of the large intestine, the caecum and appendix, are very reduced in man compared with some herbivorous mammals and have little importance. The colon has no digestive function but plays an important role in the conservation of water. In man, about 400 cm$^3$ of fluid passes from the ileum into the colon each day. Much of this water is reabsorbed by the mucosa leaving semi-solid cylinders of waste food, the **faeces**, which pass into the rectum for storage. The faeces are coloured brown by the bile pigments, breakdown products of haemoglobin.

Stretching of the rectum brings about a reflex action in which muscles of both colon and rectum contract and evacuate the faeces. Man has some voluntary control over this evacuation by the action of the external anal sphincter. This evacuation is known as **egestion** and it refers to the removal of food which has never been part of the body of the animal. Strictly speaking, the bile pigments are excretory products as they have been involved in the body's metabolism.

### 3.4.9 Digestion and absorption in herbivorous mammals

Herbivorous mammals have a problem since the nutrients in plant food are contained within cells surrounded by thick cellulose walls. Some form of mechanical breakdown of plant tissue is needed, for example, by chewing with teeth or rasping with a radula.

The enzyme cellulase is required to break down cellulose. Mammals, however, do not produce this enzyme; but some have a symbiotic relationship with bacteria living in their gut. These bacteria produce cellulase and digest the plant food of the mammal.

In ruminants, for instance cows, sheep and deer, the bacteria are found in an enlarged, four-chambered stomach. In lagomorphs, such as rabbits and hares, they are found in an enlarged caecum and appendix. The bacteria secrete cellulase into the gut of the mammal and digestion occurs extracellularly to produce fatty acids. In the ruminants, the fatty acids are absorbed through the stomach walls.

**84   Site of cellulose digestion in ruminants**

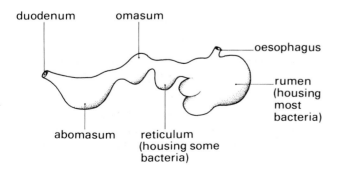

During feeding, grass passes into the rumen and reticulum where some digestion occurs (see figure 84). It is then passed back to the mouth to be chewed again. This is known as 'chewing the cud'. The finely divided food is then returned to the rumen for further bacterial digestion. The omasum acts as a strainer retaining particles which require further breakdown. The smaller particles pass into the true stomach (abomasum).

In lagomorphs, cellulose digestion occurs in the caecum and appendix (see figure 85). The products of digestion cannot be absorbed in the large intestine. These animals form a special kind of faecal pellet from the contents of the caecum and appendix. These pellets are generally produced during the day in the burrows. They are wetter, softer, larger and lighter than the firm, dark pellets produced at night. The wet pellets are eaten by the animals. This is known as **refection**. As the wet pellets pass through the alimentary canal, the products of cellulose digestion are absorbed.

**85   Site of cellulose digestion in lagomorphs**

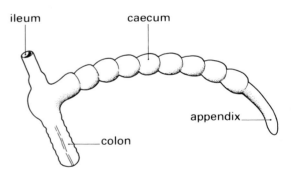

The guts of herbivores are generally longer and more complex than the guts of related carnivores which allows more time for plant food to be digested.

*SAQ 82* Why is 'chewing the cud' important?

*SAQ 83* Why must herbivorous animals have micro-organisms in their gut?

*SAQ 84* List, in order, the regions of the gut through which food passes in a sheep.

**SAQ 85** Where does cellulose digestion occur
(*a*) in a cow,
(*b*) in a rabbit?

**SAQ 86** Where are the products of cellulose digestion absorbed
(*a*) in a cow,
(*b*) in a rabbit?

## 3.4.10 The liver

The products of digestion are transported to the liver where they may be further processed. The liver, the largest gland in the body, is extremely important. Most of its functions are concerned with helping to maintain a constant internal environment in the body.

It does this in a number of ways:

(*a*) **storage** of substances present in excess, such as glycogen, vitamins and minerals;

(*b*) **synthesis** of substances required by the body, such as fatty acids, plasma proteins, non-essential amino acids and bile;

(*c*) **breakdown** of substances which are toxic or present in excess, such as amino acids, poisons, hormones and old red blood cells;

(*d*) **interconversion** of substances present in excess into those in short supply, for instance glycogen → glucose, carbohydrates → lipids.

The liver is a five-lobed structure found under the diaphragm to which it is attached by ligaments. Its gross structure is shown in figure 86.

Figure 86 shows the unusual double blood supply of the liver.

**SAQ 87** Name the two blood vessels which supply the liver. For each vessel, state what substance(s) are carried in large amounts.

*Treatment of amino acids in the liver*

Excess amino acids cannot be stored in the body and are broken down in the liver. The breakdown is a two-stage process.

(*a*) Deamination to produce a carbon compound

### 86 Structure of the liver

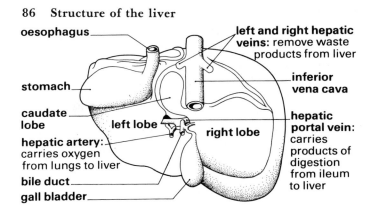

which is often converted to glucose and ammonia ($NH_3$) which is toxic. Oxygen is required for this stage.

(*b*) Detoxification of ammonia by conversion to urea in the ornithine cycle. See figure 87.

### 87 The ornithine cycle

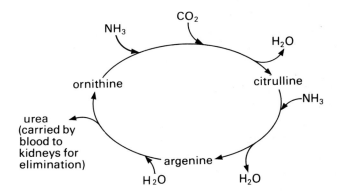

**SAQ 88** Which of the substances required for the production of urea from ammonia
(*a*) are re-usable?
(*b*) are not re-usable?

*Treatment of glucose in the liver*

The liver is responsible for helping to maintain the blood glucose level at about 0.1 g per 100 cm³. When glucose levels rise, glucose is converted to glycogen and stored in the liver. The glycogen may subsequently be broken down if glucose levels fall. Hormones from the pancreas regulate these interactions.

If glucose and glycogen levels continue to rise, glucose may be converted to fats for storage.

*Treatment of lipids in the liver*

Absorbed fats, fatty acids and glycerol pass to the liver via the blood and lymph transport systems. Much of this material then passes to the fat storage regions of the body, particularly under the skin. Some of the excess cholesterol in the diet is eliminated from the body via the bile.

### 3.4.11 Assimilation

The way in which the body uses the products of digestion is shown in figure 88. Study this carefully and then answer the following questions.

**88 Assimilation of the products of digestion**

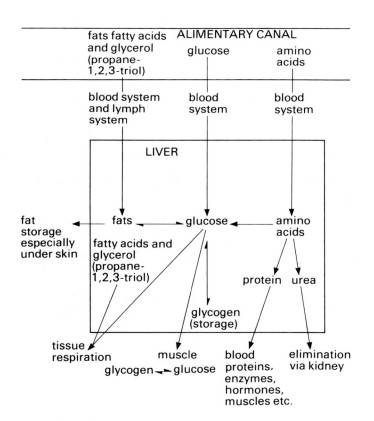

**SAQ 89** List the products absorbed from the alimentary canal into the transport systems.

**SAQ 90** State three things that might happen to absorbed glucose.

**SAQ 91** Which products of digestion are usually used in respiration?

**SAQ 92** State three things which might happen to amino acids in the liver.

Write an essay to answer the following past examination question.

Outline the parts played by the stomach, liver, small intestine and pancreas in the life of a named mammal. (London, June 1978).

Show this work to your tutor.

### 3.4.12 Summary assignment for section 3.4

**1** Draw an annotated diagram of the human gut. Label the main regions. Indicate the sites of ingestion, mechanical and chemical digestion, absorption, assimilation and egestion.

**2** Draw up a table to summarise the digestion of carbohydrates, proteins and lipids based on figure 75.

**3** Draw an annotated diagram of a section through the ileum. Indicate the main regions and add notes on the functions of each.

**4** Describe how amino acids, lipids and monosaccharides are absorbed from the small intestine into the bloodstream.

**5** Explain why digestion is a problem in herbivorous mammals and how these problems are overcome (*a*) in a sheep, and (*b*) in a rabbit.

**6** (*a*) Draw a simple diagram to show the structure and blood supply of the liver.
(*b*) List the major functions of the liver.

**7** List the products of digestion which are absorbed in the ileum and state what happens to them.

Show this work to your tutor.

Self test 11, page 93, covers section 3.4 of this unit.

# Section 4 Respiration

## 4.1 Introduction and objectives

You now know how organisms obtain their food, whether they are autotrophic like green plants, or heterotrophic like animals.

One of the most important functions of this food is to provide an energy source. Energy is required for such diverse functions as the synthesis of new materials, transport of substances within the body, movement and transmission of nerve impulses.

The process during which energy is released from food molecules in a form useable by living organisms is known as respiration. This unit deals with this process. Respiration often requires gaseous oxygen and results in the production of carbon dioxide. Details of gaseous exchange mechanisms are covered in the unit *Exchange and transport* in this series.

After completing this section, you should be able to do the following.

(a) State the importance of respiration to living organisms.

(b) Name the two main types of respiration.

(c) Describe precisely the location of reactions of respiration within living organisms.

(d) Describe the significance of the following: glycolysis, the Krebs cycle, respiratory chain, oxidative phosphorylation.

(e) Explain why ATP is required during respiration.

(f) Explain what is meant by: oxidation, reduction, redox reaction, oxidising agent, reducing agent.

(g) Explain where oxidation occurs in respiration. Explain where carboxylation occurs in respiration.

(h) Describe how ATP formation is coupled with the biochemical reactions of respiration.

(i) Name two hydrogen acceptors in cells and three electron acceptors in cells.

(j) Describe the role of hydrogen acceptors and electron acceptors in respiration.

(k) Explain the importance of vitamins to respiration.

(l) Describe the role of oxygen in respiration.

(m) Outline the structure and functions of mitochondria.

(n) List the main differences between aerobic and anaerobic respiration.

(o) Compare anaerobic respiration in plants and animals.

(p) Write equations for aerobic respiration, lactic acid fermentation and alcohol fermentation.

(q) State where proteins and fats enter the respiratory pathway.

(r) Write an equation for the respiratory quotient, and state the respiratory quotient for aerobic respiration of glucose, proteins and fats, and for the anaerobic respiration of glucose.

(s) Explain how the metabolism of mammalian muscle adapts it for vigorous exercise.

*Extension*

(a) Explain how cytochromes are involved in redox reactions.

## 4.2 Energy from food

---

**AV 5: Energy from food**

---

Respiration is the process by which all living cells obtain their energy in a readily useable form. This energy is released from their food source during a series of biochemical reactions. Each of these reactions is controlled by a specific enzyme.

The following video may be used as an introduction and as a summary to the process of respiration.

### Materials

VCR and monitor
ABAL video sequence: *Energy from food*
Worksheets

### Procedure

(*a*) Check that you have all the relevant materials for this activity.

(*b*) Check that the video cassette is set up ready to show the appropriate sequence – *Energy from food.*

(*c*) Start the VCR and stop it to complete the work sheets as instructed.

(*d*) If you do not understand anything, stop the video, rewind and study the relevant material again before consulting with your tutor.

(*e*) If possible, work through the video and worksheets with a small group and discuss the material with your fellow students.

Show this work to your tutor.

---

## 4.3 Types of respiration

Glucose is the molecule most commonly used by organisms as an energy source.

In the presence of oxygen, glucose is broken down to carbon dioxide and water. This is known as **aerobic**

**respiration**. During the process the energy which is released is transferred to ATP molecules.

$$\text{glucose} + \text{oxygen} \longrightarrow \text{carbon dioxide} + \text{water} + \text{energy}$$

$$C_6H_{12}O_6 + 6O_2 \longrightarrow 6CO_2 + 6H_2O + \text{energy}$$

In some circumstances, organisms can obtain energy from sugar in the absence of oxygen. This is known as **anaerobic respiration**.

The reactions of anaerobic respiration are different for animals and plants. They are summarised in the following equations.

*Plants (including fungi)*

$$\text{glucose} \longrightarrow \text{ethanol} + CO_2 \text{ energy}$$

$$C_6H_{12}O_6 \longrightarrow 2CH_3 \cdot CH_2 \cdot OH + 2CO_2 + \text{energy}$$

This is known as **alcoholic fermentation**.

*Animals*

$$\text{glucose} \longrightarrow \text{lactic acid} + \text{energy}$$

$$C_6H_{12}O_6 \longrightarrow 2CH_3 \cdot CH \cdot OH \cdot COOH + \text{energy}$$

This is known as **lactic fermentation**.

All living cells require energy, hence all require a source of glucose and usually also of oxygen.

The stages of aerobic respiration occur within the cytoplasm and in the mitochondria. Anaerobic respiration takes place solely in the cytoplasm.

Figure 89 summarises the main stages of respiration which you will study in more detail in following sections.

*SAQ 93* What stage is common to both aerobic and anaerobic respiration?

*SAQ 94* Name the raw materials and the products of the following stages. Give your answer in table form.
(*a*) Glycolysis      (*b*) Krebs cycle
(*c*) Respiratory chain      (*d*) Oxidative phosphorylation

*SAQ 95* What are the products of respiration
(*a*) in the absence of oxygen?
(*b*) in the presence of oxygen?

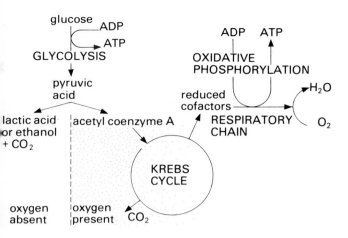

You do not need to know all the biochemical details of the sequence of reactions by which energy is released from glucose. However, there are three points which you must understand about the process. These are that it involves

(*a*) the breakdown of a 6-carbon molecule to 6 molecules of carbon dioxide;

(*b*) a series of oxidation and reduction reactions;

(*c*) a small input of energy followed by a relatively greater output of energy.

These points are covered in sections 4.4, 4.5 and 4.6 respectively.

# 4.4 Breakdown of the carbon skeleton of glucose

The 6-carbon glucose molecule is broken down to six 1-carbon molecules of $CO_2$ as shown in figure 90.

Study this diagram and compare it with figure 89.

*SAQ 96* What has happened to the carbon skeleton by the end of glycolysis?

In the conversion of a pyruvic acid molecule to acetyl coenzyme A (acetyl coA), the carbon skeleton is further broken down and a molecule of $CO_2$ is evolved. This is referred to as **decarboxylation**.

Acetyl coA then enters a cycle of reactions.

*SAQ 97* What role does the Krebs cycle play in the breakdown of the carbon skeleton?

*SAQ 98* What happens to the carbon skeleton after glycolysis in anaerobic respiration in animals and plants?

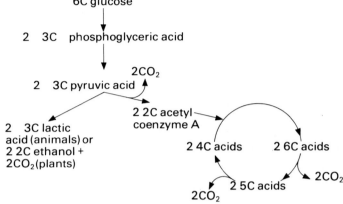

## 4.5 The oxidation of glucose

In order to understand the oxidation of glucose, it is necessary to understand something about oxidation and reduction in general.

These ideas are covered in the following programmed learning text.

### 4.5.1 Programmed learning text: Oxidation and reduction

For an explanation of how to work through this programmed text, see the section 'How to use this unit' at the front of the book.

**1** Oxidation is most often defined in simple terms as the addition of oxygen or the removal of hydrogen.

The following equation shows the oxidation of magnesium to magnesium oxide.

**Equation 1** $\quad 2Mg + O_2 \longrightarrow 2MgO$

Reduction is defined as the removal of oxygen or the addition of hydrogen.

The following equation shows the reduction of copper oxide to copper.

**Equation 2**  $CuO + C \longrightarrow Cu + CO$

**Q** What happens to the carbon in equation 2?

---

**A** It is oxidised.

**2** This illustrates a general principle which is that oxidation and reduction reactions always occur simultaneously. For this reason they are called **redox reactions** (from **red**uction/**ox**idation).

Respiration involves a series of redox reactions.

**Q** In the following two equations, say which of the reactants are (*a*) oxidised, (*b*) reduced.

**Equation 3**  $2H_2S + 3O_2 \longrightarrow 2H_2O + 2SO_2$

**Equation 4**  $2H_2 + O_2 \longrightarrow 2H_2O$

---

**A** (*a*) $H_2S$, $H_2$ (*b*) $O_2$, $O_2$

**3** A more accurate definition of oxidation and reduction involves a consideration of the electrons in the substances.

The equation for the oxidation of calcium by oxygen is shown below.

**Equation 5**  $2Ca + O_2 \longrightarrow 2CaO$

In terms of the atoms and their constituent electrons, it may be represented as in figure 91.

**91    Reaction of calcium with oxygen**

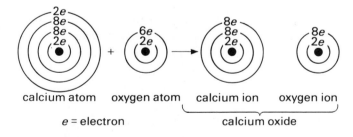

$e$ = electron  calcium oxide

Notice that a redistribution of electrons occurs.

**Q** How many electrons has
(*a*) a calcium atom?    (*b*) a calcium ion?
(*c*) an oxygen atom?    (*d*) an oxygen ion?

---

**A** (*a*) 20   (*b*) 18   (*c*) 8   (*d*) 10

**4** The calcium atom has lost two electrons and become a positively charged ion $Ca^{2+}$.

The oxygen atom has gained two electrons and become a negatively charged ion $O^{2-}$.

A more convenient way of emphasising the redistribution of electrons in this reaction involves considering only the changes in charge that occur.

**Equation 6**  $2Ca + O_2 \longrightarrow 2Ca^{2+}O^{2-}$
calcium oxide

The loss of electrons is associated with oxidation. The gain of electrons is associated with reduction. You may find it helpful to use the mnemonic **oilrig** to remember this: **oxidation is loss of electrons (oil)**, **reduction is gain (rig)**.

**Q** Study the following equation and indicate the charge distribution which occurs using equation 6 as a model.

**Equation 7a**  $2Cu + O_2 \longrightarrow 2CuO$
copper   oxygen              copper oxide

---

**A Equation 7b**  $2Cu + O_2 \longrightarrow 2Cu^{2+}O^{2-}$

**5** State which substance has been oxidised and which has been reduced. Give an explanation in terms of electrons.

---

**A** The copper atom has been oxidised because it has lost two electrons. The oxygen atom has been reduced because it has gained two electrons.

**6** Below are equations for two reactions. For each, say whether the metal atom is oxidised or reduced.

**Equation 8**
$Cu^{2+}O^{2-} + C \longrightarrow Cu + CO$
copper    carbon              copper    carbon
oxide                                    monoxide

**Equation 9**

$$2Fe^{2+} + Cl_2 \longrightarrow 2Fe^{3+} + 2Cl^-$$

iron(II) salt　　chlorine　　　　iron(III) salt　　chloride

---

**A** Equation 8 – copper is reduced.
Equation 9 – iron is oxidised.

**7** In the following equation, magnesium is oxidised to magnesium chloride by removal of electrons.

**Equation 10**　$Mg + Cl_2 \longrightarrow Mg^{2+}(Cl^-)_2$

The electrons removed are accepted by the chlorine which is said to be an **oxidising agent**.

At the same time, chlorine atoms are reduced to chloride ions with electrons donated by magnesium. Thus, magnesium is said to be a **reducing agent**.

**Q** Choose the correct alternative below.
(*a*) Oxidising agents **donate/accept** electrons.
(*b*) Reducing agents **donate/accept** electrons.

---

**A** (*a*) accept (*b*) donate

**8** Substances, that is oxidising and reducing agents, vary in their tendency to attract electrons relative to one another. If a substance B has a greater tendency to attract electrons than a substance A, B will act as an oxidising agent and accept electrons from A. If another substance C is present whose tendency to attract electrons is greater than that of A and B, electrons will then pass onto C.

Consider the following sequence of oxidising and reducing agents arranged in order, from left to right, of increasing tendency to attract electrons.

$$\textbf{V} \quad \textbf{W} \quad \textbf{X} \quad \textbf{Y} \quad \textbf{Z}$$

increasing tendency to attract electrons $\longrightarrow$

**Q** (*a*) Which is the strongest oxidising agent?
(*b*) State the order in which you would expect electrons to move starting from X.

---

**A** (*a*) Z (*b*) X $\longrightarrow$ Y $\longrightarrow$ Z

**9** Redox reactions occur in living cells. The oxidation of organic compounds is an exergonic (exothermic) process, that is energy is released.

**Q** Choose the correct alternative below.
The exergonic process whereby glucose is broken down during respiration is important in energy relations of the cell because it involves an energy **input/release**.

---

**A** Release

Now answer the following questions. If you do not get full marks you should go through the programmed learning text again or ask for help from your tutor.

*Programmed learning text post-test*

**1** Fill in the blanks.
Oxidation is the (i) _____ of oxygen,
(ii) _____ of hydrogen,
(iii) _____ of electrons,

**2** Which of the following reactions results in energy release?
(i) Removal of electrons from an organic compound.
(ii) Addition of electrons to an organic compound.
Indicate the correct alternative.

**3** (i) An oxidising agent donates/accepts electrons.
(ii) A reducing agent donates/accepts electrons.
Indicate the correct alternative.

**4** In each of these equations identify
(i) which two substances have been reduced;
(ii) which two substances have been oxidised.

$$2Na + Cl_2 \rightarrow 2NaCl$$
$$CuO + C \rightarrow Cu + CO$$

**5** The diagram below indicates some organic molecules arranged in increasing order from left to right of their tendency to accept electrons

$$\textbf{A} \quad \textbf{B} \quad \textbf{C} \quad \textbf{D} \quad \textbf{E} \quad \textbf{F}$$

(i) Indicate the direction of electron flow. It is from **A** to **F** or **F** to **A**?
(ii) Which of the above molecules is the strongest

reducing agent?
oxidising agent?

Check your answers with those on page 111.

## 4.5.2 Respiration as a series of redox reactions

During the breakdown of the carbon skeleton, at various points pairs of hydrogen atoms are removed. This removal is called **dehydrogenation** and is controlled by enzymes called oxidoreductases (**dehydrogenases**). Each of these reactions involves the oxidation of the carbon skeleton.

The pairs of hydrogen atoms which are removed are picked up by coenzymes which act as **hydrogen carrier molecules** such as NAD (nicotinamide adenine dinucleotide) which consequently becomes reduced to $NADH_2$.

The NAD molecules act in close association with the dehydrogenase enzymes and hence are termed **coenzymes**.

Molecules called flavoproteins (FP) are also involved in the redox reactions of respiration as hydrogen carriers. Thus FP becomes reduced to $FPH_2$.

You may remember that nicotinic acid, important for the formation of NAD, is one of the vitamins of the B group. Riboflavin, important in FP, is vitamin $B_2$.

## 4.5.3 The respiratory chain – part 1

Before starting this section, refer back to figure 90 to remind yourself of the point at which the respiratory chain reactions occur in relation to the rest of the reactions of respiration.

Reduced NAD ($NADH_2$) and reduced FP ($FPH_2$) transfer their hydrogen to other hydrogen carrier molecules in the respiratory chain which consequently become reduced. The $NADH_2$ and $FPH_2$ themselves are oxidised and are therefore available again to be used in further reduction reactions.

Thus, a short chain of redox reactions is set up. FP enters the chain at a later point than NAD. This is illustrated in figure 92.

**92  The respiratory chain – part 1**

$coQ$ = coenzyme Q

The reduced coenzyme Q is then oxidised. The hydrogen ions released pass into the mitochondrial matrix while the electrons pass into a further sequence of redox reactions.

$$coQH_2 \longrightarrow coQ + 2H^+ + 2e^-$$

The molecules involved in this further sequence of redox reactions are known as **cytochromes**.

## 4.5.4 Extension: Cytochromes

Cytochromes are pigments and similar in structure to chlorophyll and haemoglobin (the respiratory pigment in blood. They consist of a protein bound to an iron-containing group.

The iron in the molecule can pick up and release electrons. Thus, the molecule is important in redox reactions in which the following changes occur.

$$Fe^{3+} + e^- \rightleftharpoons Fe^{2+}$$

oxidised     reduced
state       state

There are several cytochromes involved in respiration. They are named alphabetically, *a, b* and *c*. If the cytochromes are arranged in order of increasing tendency to accept electrons, the following sequence emerges.

cyt *b*, cyt *c*, cyt *a*,

increasing tendency to accept electrons

*SAQ 99* In which direction would you expect electrons to flow?

$$b \longrightarrow a \text{ or } a \longrightarrow b$$

Cytochrome molecules are also involved in electron transport in photosynthesis.

# 4.5.5 The respiratory chain – part 2

Once reduced, coQ is oxidised again, the electrons are transferred to a series of cytochrome molecules in a short chain of redox reactions. At the end of the chain, oxygen acts as the final electron acceptor (see figure 93).

**93    The respiratory chain – part 2**

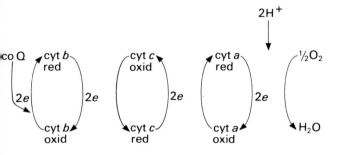

*SAQ 100* Choose the correct alternative. Oxygen is the final electron acceptor because it has the strongest/weakest tendency to accept electrons.

In addition to accepting electrons, the oxygen molecule picks up hydrogen ions from the cellular matrix. As a result, water is formed. This reaction is controlled by the enzyme cytochrome *c*: oxygen oxidoreductase (cytochrome oxidase).

*SAQ 101* In this reaction in which oxygen accepts electrons and hydrogen ions, is the oxygen molecule being oxidised or reduced?

At certain points during the chain of redox reactions, considerable amounts of energy are released during the oxidation phases. Remember that oxidation is an exergonic (exothermic) reaction. The energy released is coupled with the synthesis of ATP molecules as described in the next section.

---

## Practical P: Investigating the redox reactions of respiration

---

You have learnt that at certain stages during the breakdown of glucose during respiration, oxidation occurs by removal of hydrogen atoms. These hydrogen atoms are normally picked up by NAD.

The dye, methylene blue, acts as a hydrogen acceptor and, once it has picked up hydrogen atoms, it becomes colourless. In this practical, you will investigate the effects of yeast on methylene blue in varying conditions.

### Materials

10 cm³ 20% yeast suspension (prepared two days in advance), distilled water, 10 cm³ 0.005% methylene blue solution, 5 cm³ 1% glucose solution, test-tube rack and four test-tubes with tight-fitting bungs, test-tube holder, Bunsen burner, 2 × 2 cm³ pipettes, water-bath at 35–45 °C, labels or spirit marker

### Procedure

(a) Boil about 3 cm³ of yeast suspension in a test-tube for 1 min and then allow to cool.

(b) Label three test-tubes **A, B** and **C**.

(c) Add the substances to the test-tubes as shown in figure 94.

**94    Making up test-tubes for practical P**

| | Test-tube | |
| A | B | C |
| --- | --- | --- |
| 2 cm³ boiled yeast | 2 cm³ unboiled yeast | 2 cm³ unboiled yeast |
| 2 cm³ distilled water | 2 cm³ distilled water | 2 cm³ 1% glucose solution |

(d) Place the tubes in a water-bath at 35–45 °C for 30 min.

(e) Add 2 cm³ methylene blue to each tube and fit the bungs. Shake thoroughly for about 20 s. Return the tubes to the water-bath and do not shake any more. Note the time.

(f) Observe the tubes to see how long it takes for the blue colour to disappear. A thin film of blue colour at the surface of the tube may be ignored.

(g) The investigation may be repeated by removing the bungs from the tubes, to expose the solutions to the air, and gently shaking until the blue colour returns.

(h) Record your results in tabular form.

**Discussion of results**

**1** Present your results in a suitable graphical form.

**2** Explain these results as fully and as concisely as possible.

**3** Why were you told not to shake the tubes any more in procedure point (e)?

Show this work to your tutor.

## 4.6 The energetics of respiration

ATP and ADP are important molecules in energy exchange reactions within living organisms. ADP can be converted to ATP and vice versa accompanied by considerable gain or loss of energy, according to the following equation:

$$ADP + P_i + energy \rightleftharpoons ATP$$
($P_i$ represents inorganic phosphate)

ATP is a relatively small, soluble molecule which can easily be transported from the site of energy release (respiration) to the sites where energy is required, such as ribosomes for protein synthesis, membranes for active transport of molecules and muscles for movement.

During the flow of electrons along the respiratory chain from NAD to oxygen, sufficient energy is released to allow the synthesis of three ATP molecules. The flow of electrons from flavoproteins to oxygen results in the synthesis of two ATP molecules (see figure 95).

**95  Coupling of energy release during redox reactions of the respiratory chain to ATP synthesis**

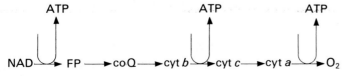

This way of forming ATP is known as **oxidative phosphorylation**. A small amount of ATP is synthesised during glycolysis, which does not involve the respiratory chain and oxidative phosphorylation.

Although a glucose molecule contains considerable quantities of energy, it is relatively unreactive. In order for it to release energy in the sequence of reactions of respiration, it must be supplied with an extra source of activation energy. Therefore, in the initial stages of glycolysis, ATP is actually required.

The sites of ATP production and use are shown in figure 96. The net production of ATP molecules per glucose molecule respired is 38 for aerobic respiration and two for anaerobic respiration.

**SAQ 102** How is the equation

$$C_6H_{12}O_6 + 6O_2 \longrightarrow 6CO_2 + 6H_2O + energy$$

misleading as a summary of the respiration of glucose?

## 4.7 Mitochondria

The structure of mitochondria was examined in the *Cells and the origin of life* unit. This should be revised or read in an alternative biology text-book. Particularly recommended is *Patterns of Biology* by D. Harrison.

The main features of mitochondrial structure are outlined in figure 97.

The functions of the various regions of the mito-chondria are outlined as follows.

**Outer membrane** – permeable to ADP, ATP, $NADH_2$, NAD, acetyl coA, fatty acids, $CO_2$ and $O_2$.

**Inner membrane** – bears the enzymes and carrier molecules of the respiratory chain and oxidative phosphorylation. Stalked granules on the cristae are associated with ATP-synthesising enzymes.

**Matrix** – contains enzymes of the Krebs cycle. Possibly loosely associated with the inner membrane. Also contains enzymes of fatty acid breakdown.

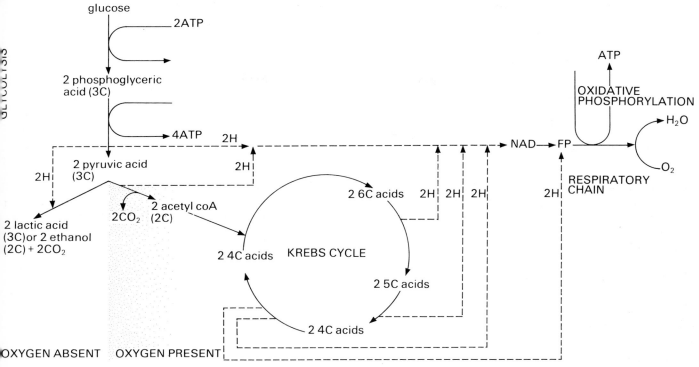

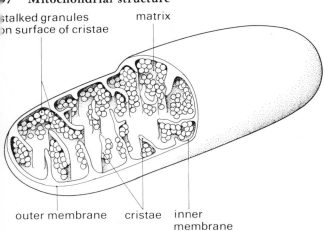

Cells requiring large amounts of energy, such as those involved in active secretion in the intestine wall and the kidney and muscle cells involved in movement, contain larger numbers of mitochondria than other cells. Also, their cristae are more numerous and more closely packed and often cross from wall to wall of the mitochondria. This provides an increased surface area for the components of the energy-releasing mechanism.

## 97  Mitochondrial structure

Write an essay to answer the following past examination question. Refer back to section 2.7.5 and figure 50.

Describe the structure and occurrence of chloroplasts and mitochondria and discuss their function. (London, June 1979)

*Reference*

*Chloroplasts and Mitochondria* by Michael Tribe & Peter Whittaker, Studies in Biology No. 31.

## 4.8 Respiratory substrates other than glucose

Many organisms use substrates other than glucose as energy sources. Carbohydrates, such as starch in plants and glycogen in animals, can be converted by phosphorylase to glucose-6-phosphate (see figure 98), and then be respired.

Lipids are first digested to fatty acids and glycerol. Fatty acids are converted to acetyl coenzyme A, and enter the Krebs cycle. Glycerol is converted to phosphoglyceraldehyde and enters glycolysis.

The amino acids from protein digestion can also be respired. Nitrogen is first removed during deamination. The remaining carbon-containing residue can then enter the respiratory pathway via pyruvic acid, acetyl coA or oxaloacetic acid (one of the acids of the Krebs cycle).

Because these other substrates easily enter the respiratory pathways, organisms that feed mainly on proteins (for instance some carnivores) or fats and oils (for instance some birds) need no special respiratory mechanisms to obtain energy.

The role of these substrates in respiration is summarised in figure 98.

**98 The role of various respiratory substrates in respiration**

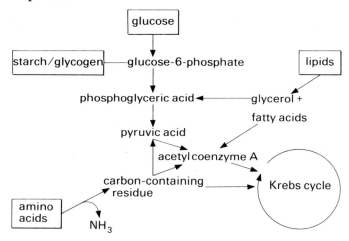

**4.8.1 The respiratory quotient**

The ratio of carbon dioxide produced to oxygen consumed in an organism or tissue is called the respiratory quotient (RQ).

$$RQ = \frac{\text{amount of } CO_2 \text{ produced}}{\text{amount of } O_2 \text{ used}}$$

The importance of the RQ is that it gives an **indication** of the substrates being used in respiration. For example, the aerobic respiration of a carbohydrate such as glucose is

$$C_6H_{12}O_6 + 6O_2 \longrightarrow 6CO_2 + 6H_2O$$

$$RQ = \frac{6 \text{ vols. } CO_2}{6 \text{ vols. } O_2} = 1$$

The RQ for lipids varies because of the differences in their composition. Respiration of the lipid tripalmatin occurs according to the following equation:

$$2C_{51}H_{98}O_6 + 145O_2 \longrightarrow 102CO_2 + 98H_2O$$

***SAQ 103*** Calculate the RQ for tripalmatin.

Lipids in general, usually have an RQ of about 0.7.

Proteins vary considerably in composition and have RQ values between 0.5 and 0.8.

As well as giving an indication of the substrates used in respiration, RQs can indicate whether respiration is aerobic or anaerobic. The anaerobic respiration of glucose is shown in the following equation:

$$C_6H_{12}O_6 \longrightarrow 2CH_3CH_2OH + 2CO_2$$

Its RQ is calculated with the following results:

$$RQ = \frac{\text{vol. } CO_2}{\text{vol. } O_2} = \frac{2}{0} = \infty$$

In this case, the RQ is higher than 1.0, and is an indication that anaerobic respiration is occurring.

Both aerobic and anaerobic respiration may occur together.

## 4.9 Anaerobic respiration in mammalian muscle

During exercise, muscle cells must make available large amounts of energy in the form of ATP molecules. At first, the breathing rate increases to supply more oxygen to the tissues for increased aerobic respiration. However, with sustained exercise oxygen cannot get to the tissues fast enough to permit sufficient energy release by aerobic respiration.

Under these conditions, mammalian muscles can respire anaerobically. The mechanism of anaerobic respiration used is lactic fermentation.

The lactic acid accumulates during exercise. Too

much lactic acid in the body causes cramp or muscle fatigue. This occurs because the lactic acid affects the nerve–muscle junction and prevents the muscle responding.

Following exercise, panting occurs. This results in the uptake of large amounts of oxygen which is used to oxidise the lactic acid. This oxidation releases more energy and converts lactic acid to carbon dioxide and water. The extra oxygen required to oxidise the accumulated lactic acid is called the **oxygen debt**.

Energy is made available to active mammalian muscle in another way. If glycolysis and oxidative phosphorylation fail to provide sufficient ATP, ADP can be phosphorylated by **phosphocreatine**. This exists in high concentrations in mammalian muscle tissue.

phosphocreatine + ADP $\longrightarrow$ creatine + ATP

In the recovery period following exercise, phospho-creatine is regenerated from free creatine at the expense of ATP formed by oxidative phosphory-lation.

*SAQ 104* Which of the following would you expect to find in high concentration in muscle tissue:
    lactic acid, $CO_2$, phosphocreatine, creatine
(*a*) during violent exercise?
(*b*) at the end of the recovery period?

*SAQ 105* Choose the best definition for the term oxygen debt.
(*a*) The oxygen required during violent exercise to provide the extra energy needed.
(*b*) The extra oxygen required to oxidise lactic acid following anaerobic respiration in mammalian muscle.
(*c*) The oxygen required to prevent respiration in active muscle from becoming anaerobic.

You may find it useful to work through the video programme *Energy from food* again now.

Write an essay to answer the following examination question.

What are the **main** differences between aerobic and anaerobic respiration? In what circumstances would

you expect anaerobic respiration of glucose to occur in (*a*) mammals, (*b*) a green plant, (*c*) yeast? (London, June 1974).

Show this to your tutor.

## 4.10 Summary assignment for section 4

**1** Define respiration.

**2** Explain the importance of respiration.

**3** Tabulate the differences between aerobic and anae-robic respiration (see self test 12, question 11).

**4** Draw a diagram summarising the biochemistry of respiration based on figure 96.

**5** Draw a diagram showing how carbohydrate, fats and proteins can enter into the sequence of bio-chemical reactions of respiration.

**6** Explain the term respiratory quotient and comment on its usefulness.

**7** Explain how the oxygen debt occurs and how it is paid off.

Show this work to your tutor.

Self test 12, page 95, covers section 4 of this unit.

# Section 5 Energy flow and nutrient cycling in ecosystems

## 5.1 Introduction and objectives

In this section you will study the way in which energy and materials move through ecosystems. In particular, the similarities and differences between the two processes will be emphasised.

At the end of this section you should be able to do the following.

(a) State how energy is obtained, utilised and lost by each trophic level.

(b) Compare the energy content of each trophic level.

(c) Construct a diagram to illustrate the flow of energy through an ecosystem.

(d) Construct diagrams to illustrate pyramids of energy, number and biomass.

(e) List the advantages and disadvantages of using pyramids of energy, number and biomass to compare ecosystems.

(f) Construct a diagram to show the cycling of nutrients through an ecosystem.

## 5.2 Energy and ecosystems

In section 1 you saw how organisms were dependent on one another and on the environment for their energy.

Section 2 demonstrated how autotrophs (green plants) obtain their energy from the abiotic environment, while in section 3 it was seen that heterotrophs (animals and non-green plants) obtained energy from other organisms, either alive or dead. The last section, study of respiration, introduced you to the process whereby the energy in organic molecules in living organisms is released in a form which is usable for their life processes.

*SAQ 106* Name four life processes which depend upon energy release during respiration.

*SAQ 107* (a) What is the ultimate source of energy for all ecosystems?
(b) Through which trophic level does this energy enter food chains?
(c) What is the process by which this energy is 'trapped' or 'fixed' by living organisms?

*SAQ 108* To which trophic level do the following belong?
(a) A moth caterpillar feeding on oak leaves
(b) A chiffchaff feeding on the caterpillar
(c) A tawny owl feeding on the chiffchaff
(d) The oak tree

### 5.2.1 Energy flow through the producer trophic level

Energy enters this trophic level as light and is converted to chemical energy during photosynthesis. The total amount converted is called the **gross primary production**, and is used for two main purposes. Some is required for the synthesis of new plant material in growth, repair and reproduction. This results in an increase in the mass of living material or biomass, and is called the **net primary production**. The rest of the energy is used in activities such as secretion, internal transport and movement, and is collectively called **respiration**.

At the same time, energy may leave this trophic level in a number of ways as follows.

(a) No energy transformations are 100% efficient, because some heat energy released from food during respiration results in the synthesis of ATP molecules. The heat released does have an important role to play in maintaining body temperature, especially in homeothermic ('warm-blooded') animals, namely mammals and birds. However, much of the heat is lost to the environment.

Respiration energy includes this heat and the energy converted to ATP.

(b) A plant may be eaten by a herbivore.

(c) Structures produced by the plant and containing energy may be lost, for instance the shedding of leaves, bark and floral parts, the production of structures for reproduction and dispersal, such as fruits, and the production of seeds and spores which fail to germinate.

(d) When the plant dies, its remains contain a considerable amount of energy.

The remains of dead plants and the structures mentioned in (c) above, represent a loss of energy from the trophic level to the environment. However, this energy does not remain 'locked up' in these structures forever. They may be broken down by the decomposer organisms, which are able to utilise the energy contained in them for their own life processes.

Figure 99 summarises energy flow through the plant community or producer trophic level.

**99   Energy flow through the producer trophic level**

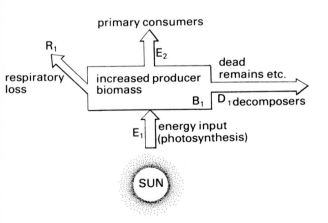

It is important to realise that energy is neither created nor destroyed. In other words, using the symbols in figure 99:

$$E_1 = B_1 + D_1 + R_1 + E_2$$

It also follows that the amount of energy available to the primary consumers ($E_2$) must be less than the amount originally 'fixed' by the producers ($E_1$) because some is lost in respiration ($R_1$), and some in the dead remains and non-living structures produced by the plants ($D_1$).

In other words, $E_2 < E_1$ because

$$E_2 = E_1 - (B_1 + D_1 + R_1)$$

**SAQ 109** Considering the total amount of energy available for growth etc., of producers and primary consumers, would you expect the total mass of living material of primary consumers to be (a) greater than, (b) less than, (c) equal to the total mass of living material of the producers?

The term **biomass** is usually used to refer to the total mass of living material.

### 5.2.2 Energy flow through the consumer trophic levels

Energy flow through the consumer trophic levels can be considered in the same way as that for the producer level.

Energy flows into the primary consumer trophic level in the form of plant material eaten ($E_2$). Some of the energy contributes to increased primary consumer biomass ($B_2$), some is lost to the environment as heat ($R_2$) and some passes to the decomposers ($D_2$) in the form of dead bodies, undigested material in faeces and excretory products.

Finally, some passes on to the secondary consumer level as herbivores eaten by carnivores.

The energy transfer between successive trophic levels is estimated to be about 10% of that entering each level. It is usually much lower.

Figure 100 summarises energy flow through the primary consumer trophic level.

**SAQ 110** Considering the total amount of energy available for growth etc., of primary and secondary consumers, would you expect the biomass of the secondary consumers to be (a) greater than, (b) less than, (c) equal to the biomass of the primary consumers?

## 100   Energy flow through the primary consumer trophic level

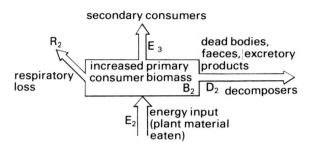

### 5.2.3 Energy flow through the whole ecosystem

It is now possible to build up a picture of energy flow through the whole ecosystem. This is shown in figure 101.

## 101   Energy flow through an ecosystem

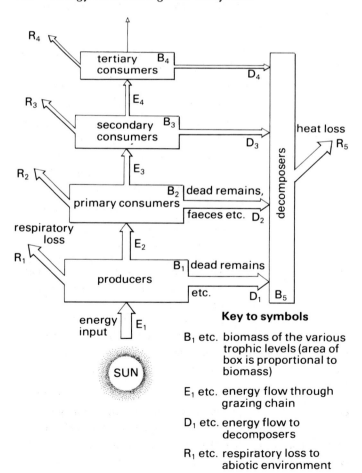

**Key to symbols**

$B_1$ etc. biomass of the various trophic levels (area of box is proportional to biomass)

$E_1$ etc. energy flow through grazing chain

$D_1$ etc. energy flow to decomposers

$R_1$ etc. respiratory loss to abiotic environment

Study the figure carefully and then answer the questions below.

*SAQ 111* (*a*) Energy enters food chains in the form of light. In what form does it eventually return to the abiotic environment?

(*b*) If the amount of light 'fixed' by green plants (energy input to producers, $E_1$) were increased, what would be the effect on the biomass of the secondary consumers?

(*c*) What would be the effect (on the biomasses of the primary consumers and producers) of removing all the carnivores from the ecosystem?

(*d*) Would you describe the movement of energy through an ecosystem as (i) a cyclical process in which the energy may be reused; (ii) a directional flow?

(*e*) What limits the number of trophic levels in an ecosystem?

## 5.3 Pyramids of energy

An alternative way of illustrating the energy levels of the different trophic levels is by means of a diagram known as a **pyramid of energy**, as shown in figure 102.

Each level of the pyramid represents a trophic level.

## 102   A pyramid of energy

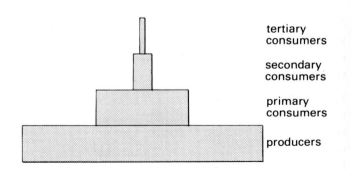

The horizontal length of each level is proportional to the total amount of energy in joules utilised by the organisms of that level per square metre per year.

## 5.4 Pyramids of number

Ecosystems are sometimes compared by means of **pyramids of number** in which the horizontal length of each level is proportional to the total number of organisms at each level. Figure 103 is an example of a pyramid of numbers.

This general trend for the numbers of organisms to decrease along food chains resulting in pyramids of numbers within an ecosystem was first recognised by Charles Elton in 1927.

However, not all ecosystems produce such perfect pyramids of numbers. For instance, exceptions occur when the producers are very large relative to the primary consumers. One oak tree can provide food for a very large number of animals.

**103   Numbers of organisms occupying each trophic level of a grassland ecosystem**

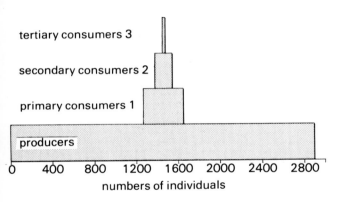

*SAQ 112* Make a diagram to show the kind of shape you might expect to obtain for an oak wood ecosystem pyramid of numbers. (Up to tertiary consumer level.)

Host–parasite food chains produce another exception to the 'classical' pyramid of numbers shape. *Ichneumon* flies (actually a kind of wasp-like insect rather than fly) lay their eggs in the bodies of the larvae of butterflies and moths. The larvae of *Ichneumon* live as parasites inside the caterpillar. However, the *Ichneumon* larvae may themselves be parasitised by the larvae of another group of wasp-like insects. A parasite which parasitises another parasite is called a **hyperparasite**.

*SAQ 113* Make a diagram to show the shape of the pyramid of numbers you might expect to obtain for this host → parasite → hyperparasite food chain.

## 5.5 Pyramids of biomass

To overcome the inadequacies of pyramids of number with respect to the size of organisms and to parasitic relationships, **pyramids of biomass** were introduced.

One way of estimating biomass is to measure the **dry mass** of a sample of organisms, calculating a mean value and multiplying by the total number of individuals in the population. The dry mass of an organism is its **fresh mass** minus the mass of its water content.

*SAQ 114* What might you expect a pyramid of biomass to look like for the oak wood and host–parasite ecosystems?

However, there are still inadequacies in the use of pyramids of biomass. One such pyramid constructed for producers and primary consumers in the English Channel appears as shown in figure 104.

**104   A pyramid of biomass for the English Channel**

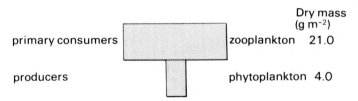

In order to explain this apparently impossible result, it is necessary to consider how the figures were obtained.

As figure 104 shows, pyramids of biomass are constructed from data expressed as mass per unit area (such as grams per square metre). In both cases, measurements are taken at a **particular point in time**.

In the case of the English Channel study, the sample from which the figures were calculated would have taken only a few minutes to collect. The information obtained from the sample thus represents a 'snapshot' – a picture of the ecosystem frozen at a particular moment in time.

If the ecosystem is studied over a longer period, an explanation of the apparently impossible figures emerges. The microscopic algae that make up the phytoplankton community have a very short life-span of only a few days, during which they carry out photosynthesis, grow and reproduce themselves. In the space of a year, a single organism could give rise to many millions of offspring. On the other hand, the small animals of the zooplankton community live rather longer and produce fewer generations of offspring in a year. Thus, the total biomass of all the producers that existed over a period of time (a year, say) would be greater than the total biomass of all the primary consumers that existed in the same period, although **at any one moment** the biomass of the consumers may be greater.

If, instead of measuring the biomass of each trophic level in the Channel, the energy utilised per square metre per year was calculated, the classical pyramid shape would be obtained.

*SAQ 115* (*a*) Which would be the best way of comparing two ecosystems? By pyramids of number, by pyramids of biomass or by pyramids of energy? (*b*) Explain your answer.

The major disadvantage of the use of pyramids of energy is that they involve the accumulation of a considerable amount of data over a long period of time, which involves much time and effort. However, this must be weighed against the benefits provided by the more useful information.

## 5.6 Cycling of nutrients in ecosystems

In addition to energy, living organisms require nutrients, both inorganic and organic. These nutrients are obtained either from other organisms or from the abiotic environment. Figure 105 summarises the dependence of organisms on each other and the abiotic environment for their supplies of nutrients.

The microorganisms involved in the decomposition stage were referred to in section 2.6.2. Further

**105   Cycling of nutrients within an ecosystem**

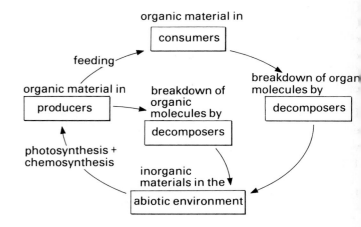

details of nutrient cycling can be found in the *Ecology* unit of this series.

*SAQ 116* How does the movement of energy and nutrients differ within an ecosystem?

Answer the following past examination questions.

The pyramid of biomass (as $g^{-2}$) shown in figure 106 is based on five sets of data from a seashore environment in the fucoid (brown algae) zone.

**106   Pyramid of biomass 1**

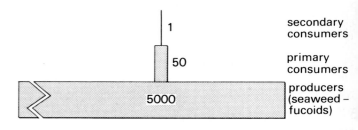

In contrast the pyramid of biomass shown in figure 107 was obtained from the same number of sets of data from a meadow.

**107   Pyramid of biomass 2**

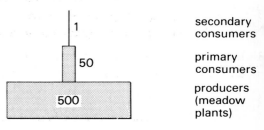

(*a*) Why do biomass plots take this pyramid shape? (5 marks)
(*b*) What is the difference in efficiency of conversion of biomass between producer and secondary consumer in the two examples given? (4 marks)
(*c*) How do you account for the differences in efficiency? (2 marks)

In five rocky pools near the low tide mark, similar estimations of biomass were made and the net result was a pyramid of apparent inverse form (see figure 108).

**108   Pyramid of apparent inverse form**

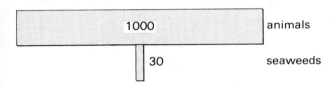

(*d*) How do you account for this inverse form? (3 marks)
(*e*) If a change in the shape of the shore line brought such a pool up into the splash zone, what change in the pyramid would be expected to occur? (3 marks)
(*f*) Make a plot of the pH changes about the neutral point that would be likely to occur in a rock pool over a 24 h period. Explain the shape of your curve. (6 marks)

(A.E.B. November 1977)          TOTAL 23 marks

## 5.7 Summary assignment for section 5

**1** Construct a diagram to show the flow of energy through an ecosystem, based on figure 101.

**2** Explain what limits the number of trophic levels in an ecosystem.

**3** Explain pyramids of energy, number and biomass. Note the advantages and disadvantages of each.

**4** Construct a diagram to show the cycling of nutrients within an ecosystem based on figure 105.

**5** State the major difference between the movement of energy and nutrients within an excosystem.

Show this work to your tutor.

Self test 13, page 96, covers section 5 of this unit.

# Section 6  Self tests

## Self test 1

**1** List three abiotic factors which might affect the distribution of organisms in a pond ecosystem.

**2** List three biotic factors which might affect the distribution of organisms in an oak wood ecosystem.

**3** Explain the following terms and give two examples of each.
(*a*) Autotrophic organism
(*b*) Heterotrophic organism
(*c*) Saprophytic organism

**4** Explain the similarities and differences between the following pairs of terms.
(*a*) Detritivore and scavenger
(*b*) Population and community
(*c*) Grazing and decay food chains

**5** Give two examples of food chains, one from a pond ecosystem and one from a woodland ecosystem. Each should contain at least three links.

**6** Give four examples of ecosystems in addition to ponds and oak woods.

**7** For each of the following organisms, state whether it is a consumer or producer.
(*a*) Cow        (*d*) Mushroom
(*b*) Man        (*e*) Dandelion
(*c*) Beech tree  (*f*) Bread mould

**8** Define the following terms:
nutrition, energy, trophic level, food web, omnivore.

**9** Give two examples of inorganic nutrients and three examples of organic nutrients required by living organisms.

## Self test 2

**1** Who was the first scientist to show that injured air required light and green vegetation to be restored?

**2** What contribution did (*a*) van Helmont and (*b*) Priestley make to an understanding of photosynthesis?

**3** Write an equation for photosynthesis
(*a*) as it was understood after the work of de Saussure;
(*b*) as it is understood today.

## Self test 3

Use information from figure 109, to answer questions **1–3**.

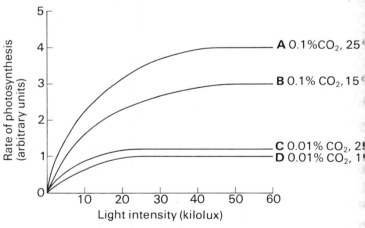

**109    Rate of photosynthesis and interaction of factors**

Choose the best answer for questions **1–4**.

**1** Increase in $CO_2$ concentration
(*a*) has no effect on photosynthetic rate.
(*b*) increases photosynthetic rate.
(*c*) decreases photosynthetic rate.
(*d*) causes a greater increase in photosynthetic rate at higher light intensities.
(*e*) causes a greater increase in photosynthetic rate at lower light intensities.

**2** Increase in light intensity
*a*) causes an increase in photosynthetic rate at higher temperatures only.
*b*) causes an increase in photosynthetic rate at lower temperatures only.
*c*) always causes an increase in photosynthetic rate.
*d*) causes an increase in photosynthetic rate at low $CO_2$ levels only.
*e*) causes an increase in photosynthetic rate at high $CO_2$ levels only.

**3** Temperature
*a*) has no effect on photosynthesis.
*b*) causes a significant increase in photosynthetic rate at all times.
*c*) causes the greatest increase in photosynthetic rate at high light intensities and low $CO_2$ levels.
*d*) causes the greatest increase in photosynthetic rate at high light intensities and high $CO_2$ levels.
*e*) causes the greatest increase in photosynthetic rate at low light intensities and high $CO_2$ levels.

**4** The compensation point of a plant is
*a*) the light intensity at which the rate of sugar synthesis equals the rate of sugar utilisation.
*b*) the light intensity at which $CO_2$ evolution equals $O_2$ evolution.
*c*) reached at the same time each day for all plants.
*d*) high for shade plants and low for sun plants.
*e*) not important in terms of survival in the natural habitat.

**5** Draw an action spectrum for photosynthesis and explain what it shows in two or three sentences.

**6** (*a*) What is an absorption spectrum?
*b*) How do the absorption spectra for chlorophyll and the accessory pigments differ?

**7** Name one type of plant in which accessory pigments are important. Explain why they are so important.

**8** Figure 110 shows the effects of light and temperature on photosynthesis. Explain how this led Blackman to postulate that photosynthesis is a two-stage process.

**110   Effects of light and temperature on photosynthesis**

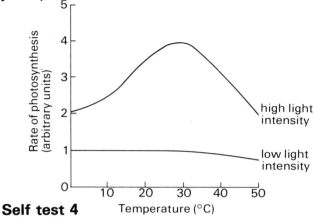

**Self test 4**

**1** Complete figure 111 which summarises the light-dependent phase of photosynthesis.

**111   Summary of light-dependent phase of photosynthesis**

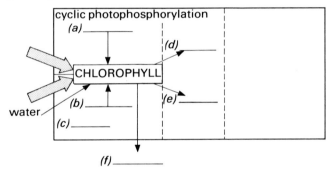

**2** Explain the following in relation to photosynthesis.
(*a*) Excited electron
(*b*) Electron flow
(*c*) Photosynthetic phosphorylation

**3** In an experiment, *Chlorella* is grown in illuminated culture tanks through which $CO_2$ is bubbled.
From your knowledge of photosynthesis, if the $O_2$ evolved was labelled ($^{18}O_2$), which of the raw materials of photosynthesis must have been labelled?

**4** (*a*) Write two equations which show the dissociation of water and hydroxyl ions.

(*b*) Explain what happens to each product of dissociation in photosynthesis.

**5** What contributions did (*a*) Robert Hill and (*b*) Arnon make to our understanding of photosynthesis?

## Self test 5

**1** What information does figure 112 give about the role of PGA and RDP in photosynthesis?

**112   Changes in amounts of photosynthetic intermediates**

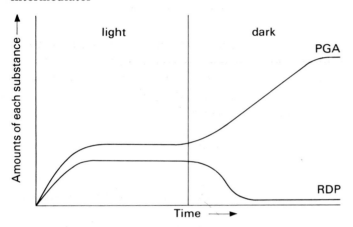

**2** Draw a simplified version of the Calvin cycle to show the following.
(*a*) Carboxylation
(*b*) Reduction
(*c*) Regeneration
(*d*) Product synthesis
(*e*) The introduction of materials from the light-dependent phase
(*f*) The number of carbon atoms in each molecule in the cycle

**3** Briefly describe how (*a*) chromatography and (*b*) autoradiography were used to work out the Calvin cycle.

**4** What is a $C_4$ plant?

## Self test 6

**1** (*a*) Write an equation for photosynthesis in sulphur bacteria.

(*b*) State the hydrogen source and name the pigment involved in light absorption.

**2** What is the source of energy for chemosynthetic bacteria?

**3** (*a*) Give a named example of a chemosynthetic bacterium.
(*b*) Write the equation for the reaction by which it obtains its energy.

**4** Why are chemosynthetic bacteria important ecologically?

## Self test 7

**1** Look at figure 113.

**113   VS of a leaf**

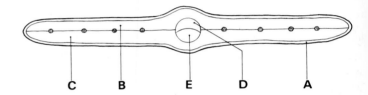

(*a*) Name the cells found in regions **A–E**.
(*b*) In which regions would you find the cells shown in figure 114.

**114   High power drawing of three leaf cells**

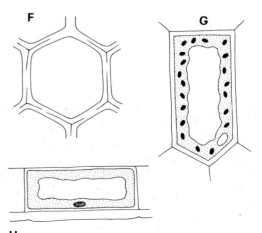

(c) Which regions would you associate with the following functions?
(i) Photosynthesis
(ii) Transport of sugars
(iii) Movement of gases
(iv) Control of gas exchange and water loss

**115 Chloroplast structure**

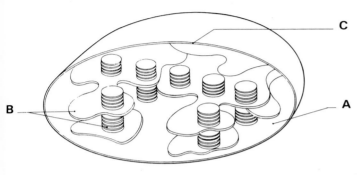

**2** Look at figure 115. Which of the labels (i)–(vi) would you associate with the labelling lines **A**, **B** and **C**?
(i) Photosynthetic pigments which absorb light
(ii) Electron carrier molecules associated with electron flow
(iii) ATP formation
(iv) $CO_2$ fixation
(v) Synthesis of sugars, amino acids, fatty acids, etc.
(vi) Control of movement of $CO_2$, water, minerals, sugars, etc.

## Self test 8

**1** Distinguish between an essential element and a trace element.

**2** Name the technique used to show the importance of minerals in plant nutrition.

**3** Name the 13 essential elements for plants.

**4** Name one element required for each of the following.
(a) Protein structure
(b) Chlorophyll structure and formation
(c) Middle lamella

(d) Cell membrane
(e) Enzyme structure or activation
(f) Reactions involving ATP

## Self test 9

**1** Give three examples of each of the following.
(a) Small particle feeders
(b) Filter feeders (other than molluscs and crustaceans)
(c) Large particle feeders
(d) Fluid feeders
State the phylum to which each belongs.

**2** The list (i)–(xv) below presents a number of items of information that refer to aspects of animal feeding. Select the items from the list that are appropriate to describe the methods of feeding of
(a) sea anemones or *Hydra*.
(b) *Daphnia*.
(c) *Arenicola* (the lugworm).
(d) mussels.

(i) The animal has a ring of tentacles that are used in feeding.
(ii) The surrounding deposit is indiscriminately ingested.
(iii) Feeding occurs more or less continuously.
(iv) Water and food enters through an inhalant siphon and leaves through an exhalant siphon.
(v) A water current is created by the beating of appendages.
(vi) Captured food is brought to the mouth by the movement of the tentacles.
(vii) Food consists of microscopic plankton living suspended in the water.
(viii) The organism has special stinging cells that trap and paralyse the prey.
(ix) Swallowing occurs periodically.
(x) The captured particles are entangled in mucus.
(xi) The organism is a deposit feeder.
(xii) A water current is produced by ciliary beating.
(xiii) The water current passes through fine, hair-like setae on the limbs.
(xiv) The organism lives within a burrow.
(xv) Prey is swallowed whole.

**3** (*a*) Name three things an *Amoeba* might eat.
(*b*) How does it detect its food?
(*c*) Name the structures it uses to catch food.
(*d*) Where is the food digested?

**4** A spider and a starfish both digest their food externally. State **four** ways in which they differ in their methods of obtaining nutrients. Arrange your answer in two columns of comparative statements.

**5** Complete the following paragraph by filling in the blanks.

Locusts possess a feeding apparatus consisting essentially of four parts arranged in layers, one below the other. A large upper lip or _____ covers the other mouthparts. An effective grinding and cutting apparatus is provided by the large _____ . Just below these lie a small pair of accessory jaws, the _____ . Attached to these are jointed _____ which are _____ in function. A shovel-like structure serves as a lower lip and is called the _____ .

**6** Make a labelled diagram of a section through a mammalian tooth. Label the following structures. Gum, enamel, dentine, pulp cavity, cement, perio-dontal membrane, alveolar bone, nerves and blood vessels

**7** (*a*) Is the skull shown in figure 116 that of a herbivore, carnivore or omnivore? Give two reasons for your choice.
(*b*) The skull shows all the teeth on one side. Write down the dental formula for this animal. Cheek teeth may be grouped together.
(*c*) What other name is given to the lower jaw?

**116   Mammal skull**

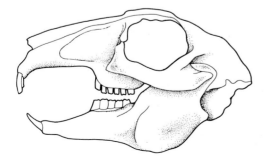

**8** Explain the difference between canine and carnassial teeth in a cat by (*a*) position, (*b*) structure, and (*c*) function.

## Self test 10

**1** Name the two major constituents of a balanced diet omitted from the following list.
Proteins, vitamins, water, carbohydrates, fibre

**2** (*a*) Name the unit used for measuring energy values of food.
(*b*) Name the term used to describe the rate at which energy is used up by the body in its resting state.
(*c*) Place the following individuals in order of increasing energy requirements.
Breast-feeding woman, coal-miner, child of six years

**3** Which of the functions listed below apply to
(*a*) carbohydrates, (*b*) lipids and (*c*) proteins?

(i) Important structural component of cell membranes
(ii) Important structural components of skin, hair, bone and muscle
(iii) Important structural components of cell walls and insect cuticles
(iv) Protection against physical damage, heat and water loss
(v) Energy source
(vi) Important in maintaining metabolism, such as controlling chemical reactions
(vii) Energy storage

**4** (*a*) Deficiency of which vitamin causes the following diseases?
(i) Scurvy
(ii) Rickets
(iii) Beriberi
(iv) Blindness
(*b*) What are the functions of thiamine, riboflavin and nicotinic acid?

**5** Name two minerals important for the following.
(*a*) Development of bones and teeth
(*b*) Constituents of body cells
(*c*) Constituents of body fluids
(*d*) Release of energy during metabolism

**6** State three reasons why water is important in the diet.

**7** Which of the following best describes the importance of fibre in the diet?
(*a*) It is an important constituent in the structure of muscle.
(*b*) It is a major factor in reducing incidence of colon tumours and appendicitis.
(*c*) It is indigestible and absorbs water in the intestine. This increases the volume of the faeces and helps prevent constipation by encouraging bowel movement.

**8** Name four important chemical components in (*a*) liver, and (*b*) bread.

**9** (*a*) What constituents are missing from fish and chips which might make it a balanced meal?
(*b*) Suggest what could be added to the meal to remedy the omissions.

**10** Figure 117 outlines the functions and tests for four main food types. Fill in the gaps to complete the table.

**117  Food tests**

| Function in body | Test reagent | Positive test result | Food type |
|---|---|---|---|
| (a) energy source | | brick-red colour | |
| (b) | sodium hydroxide copper sulphate | | protein |
| (c) energy source protection structure of cell membranes | ethanol | | |
| (d) | | dark blue-black | starch |

## Self test 11

Questions **1–8** refer to figure 118, the diagram of the alimentary canal of a mammal.

**118   Alimentary canal of a mammal**

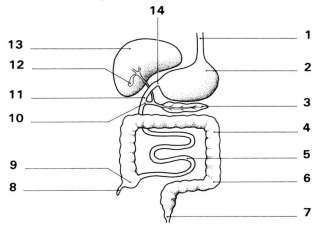

**1** The structures colon, ileum, oesophagus, caecum and duodenum are labelled respectively on the figure as
(*a*) 5, 10, 1, 7, 9?
(*b*) 6, 5, 2, 4, 8?
(*c*) 4, 5, 1, 9, 11?
(*d*) 8, 10, 5, 6, 4?
(*e*) 4, 5, 1, 7, 12?

**2** Sphincter muscles are found at
(*a*) 12 and 9.
(*b*) 2, 7 and 8.
(*c*) 8 and 10.
(*d*) 14 and 7.
(*e*) 11, 14 and 6.

**3** Water is largely absorbed in
(*a*) 2 (*b*) 5 (*c*) 6 (*d*) 9 (*e*) 4

**4** Breakdown products of haemoglobin enter the gut at
(*a*) 5 (*b*) 11 (*c*) 14 (*d*) 12 (*e*) 8

**5** Acidity is neutralised in
(*a*) 2 (*b*) 10 (*c*) 13 (*d*) 11 (*e*) 1

**6** Food is absorbed in
(*a*) 5 (*b*) 10 (*c*) 2 (*d*) 4 (*e*) 6

**7** Bile is produced in
(*a*) 2 (*b*) 3 (*c*) 13 (*d*) 12 (*e*) 10

**8** Trypsinogen is produced in
(*a*) 2 (*b*) 3 (*c*) 13 (*d*) 12 (*e*) 10

**9** Which substance is **not** a final product of digestion?
(*a*) Fructose
(*b*) Amino acids
(*c*) Maltose
(*d*) Galactose
(*e*) Glycerol

**10** Which of the following activities does **not** occur in the stomach?
(*a*) Storage of food
(*b*) Churning of food
(*c*) Chemical digestion of sugars
(*d*) Digestion of proteins
(*e*) Addition of hydrochloric acid

**11** Which substance, produced in the human digestive system, contains enzymes which can aid hydrolysis of whole protein molecules?
(*a*) Secretion of gastric glands
(*b*) Secretion of pyloric glands
(*c*) Pancreatic juice
(*d*) Saliva
(*e*) Intestinal juice

**12** Which one of the following statements best defines chemical digestion?
(*a*) The passage of food along the gut of an animal.
(*b*) The taking up by the gut of useful food substances.
(*c*) The churning of food in the stomach.
(*d*) The conversion of large or complex food molecules into smaller soluble ones.
(*e*) The breakdown of large food masses into smaller particles.

State whether the statements **13-17** are true or false.

**13** The mammalian oesophagus is an important source of digestive enzymes.

**14** Digestion of starch in a mammal is begun by the amylase in saliva.

**15** Pancreatic juice is an important source of digestive enzymes.

**16** Most digestive products are absorbed in the stomach.

**17** Fat digestion begins in the duodenum.

**18** Supply names for the labels on the transverse section of the gut shown in figure 119.

**119   TS through gut**

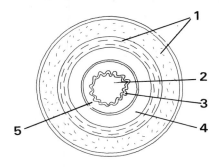

**19** Explain the difference between mechanical and chemical digestion.

**20** Define egestion.

For each of the following say which of the statements (*a*)–(*d*) are true.

**21** Herbivorous animals
(*a*) produce cellulase.
(*b*) contain bacteria in their gut.
(*c*) digest cellulose to fatty acids.
(*d*) have shorter guts than carnivores.

**22** In the liver
(*a*) oxygen is used to convert amino acids to ammonia.
(*b*) hormones are produced to control blood sugar level.
(*c*) transamination (conversion of one amino acid to another) occurs.
(*d*) vitamins are stored.

**23** Urea is produced
(*a*) by the Calvin cycle.
(*b*) by the ornithine cycle.
(*c*) in the kidney.
(*d*) from amino acids.

**24** The following are features of the ileum wall adapting it to its functions of digestion and absorption:
(a) epithelial cells with microvilli.
(b) contains digestive glands called crypts of Lieberkühn.
(c) has a smooth inner surface.
(d) possesses a rich blood supply.

**25** The liver
(a) receives oxygen via the hepatic vein.
(b) receives digested food via the hepatic portal vein.
(c) is located in the abdomen.
(d) is attached to the diaphragm.

## Self test 12

**1** What is respiration and why is it important?

**2** Link the following terms with the correct descriptions.
(a) Glycolysis
(b) Krebs cycle
(c) Oxidative phosphorylation
(d) Respiratory chain
*Descriptions*
(i) A series of redox reactions involving hydrogen and electron acceptors and during which energy is released from reduced cofactors.
(ii) A process whereby acetyl coenzyme A is broken down to carbon dioxide. ATP and reduced cofactors are synthesised.
(iii) The breakdown of glucose to pyruvic acid accompanied by a net synthesis of ATP.
(iv) The synthesis of ATP using energy released from reduced cofactors.

**3** Why is ATP required for respiration to take place?

**4** Study the following equations carefully.
(i) pyruvic acid + NAD → acetyl CoA + $NADH_2$
(ii) cytochrome $b$ + cytochrome $c$ ⟶
  ($Fe^{2+}$)    ($Fe^{3+}$)
    cytochrome $b$ + cytochrome $c$
    ($Fe^{3+}$)    ($Fe^{2+}$)
(iii) cytochrome $a$ + $\frac{1}{2}O_2$ + $2H^+$ ⟶
  ($Fe^{2+}$)
      cytochrome $a$ + $H_2O$
      ($Fe^{3+}$)

Which substances in the above equations are
(a) being oxidised?
(b) being reduced?
(c) acting as oxidising agents?
(d) acting as reducing agents?

**5** (a) Which of the equations in question 4 are redox reactions?
(b) Which of the cytochromes referred to in question 4 has the greatest tendency to accept electrons?
(c) Name one hydrogen acceptor, in addition to NAD, which is important in respiration.

**6** Explain the role of oxygen in aerobic respiration.

**7** State the precise location in the cell where the following take place.
(a) Glycolysis
(b) Krebs cycle
(c) Respiratory chain
(d) Oxidative phosphorylation
(e) Synthesis of lactic acid or ethanol from pyruvic acid

**8** Construct a diagram to show how
(a) glucose
(b) starch and glycogen
(c) lipids
(d) amino acids
may enter the glycolytic sequence and the Krebs cycle.

**9** (a) What is the respiratory quotient for the aerobic respiration of triolein?

$$C_{57}H_{104}O_6 + 80O_2 \rightarrow 57CO_2 + 52H_2O$$

(b) Suggest the most likely respiratory substrate being used if the respiratory quotient is 1.

**10** (a) Why must muscle cells switch from aerobic to anaerobic respiration during vigorous exercise?
(b) What is the end product of anaerobic respiration in muscle tissues?
(c) What is meant by the oxygen debt?
(d) What is the function of phosphocreatine in muscle tissue?

**11** Complete figure 120 as an overall summary of the processes of aerobic and anaerobic respiration.

| | Anaerobic | Aerobic |
|---|---|---|
| 1 Substrate | | |
| 2 Degree of breakdown of substrate | | |
| 3 Products | | |
| 4 Site of reactions | | |
| 5 Level of energy yield | | |
| 6 Number of ATP molecules formed per glucose molecule | | |
| 7 Stages involved include (a) glycolysis (b) reduction of pyruvic acid (c) Krebs cycle (d) respiratory chain | | |
| 8 Oxygen requirement | | |
| 9 Equation | | |

## Self test 13

1 Construct a diagram to illustrate the flow of energy through an ecosystem. Include in your diagram
(a) the point of energy input into the system,
(b) the point of energy output from the system,
(c) energy exchanges of producers, consumers (primary, secondary and tertiary) and decomposers.

2 Why are there rarely more than five trophic levels in an ecosystem?

3 With the aid of diagrams, explain what is meant by a pyramid of energy.

4 State one disadvantage for using each of the following to compare ecosystems.
(a) Pyramid of number
(b) Pyramid of biomass
(c) Pyramid of energy

5 State the major difference in the ways in which energy and nutrients move through an ecosystem.

6 Construct a diagram to show the cycling of nutrients through an ecosystem.

# Section 7 Answers to self tests

## Self test 1

**1** Any three from temperature, light availability, nature of substrate, pH, mineral content. You may think of others.

**2** Any three from food availability, predators, other organisms competing for food, other organisms dispersing seeds or causing pollination. You may think of others.

**3** (*a*) Organisms that can manufacture their own organic nutrients from inorganic nutrients. Two examples from bramble, dogwood, oak, hazel, wood anemone, primrose.
(*b*) Organisms that require a ready-made source of organic nutrients from the environment; they cannot make their own. Two examples from roe deer, chiffchaff, or any organism classified as a consumer in section 1 SAQ7.
(*c*) Organisms feeding on dead organisms or their waste. They secrete enzymes onto the material and then absorb the products of digestion. Examples are fungi e.g. crumble cap or rooting shank. Also bacteria.

**4** (*a*) Both feed on dead plants and animals. The food is taken into the gut and digested internally. Detritus feeders feed on small particles of dead matter, scavengers feed on large particles of animal origin.
(*b*) Both are groups of organisms within a particular location. A population refers to a group of organisms of the same species. A community refers to groups of organisms of different species (that is a collection of populations).
(*c*) Both represent the feeding links between organisms in an ecosystem. A grazing chain starts with living green plants. A decay chain starts with dead material.

**5** Examples of pond food chains include:
(i) broad-leaved → china mark → water → fish
pond weed       moth       boatman
                caterpillar

(ii) plankton → water snails → carnivorous beetles

Examples of oak wood food chains include:
(i) acorn → bank vole → badger
(ii) hazel → nuthatch → sparrowhawk

**6** Puddle, tropical rain forest, meadow, sandy shore

**7** (*a*) consumer      (*d*) consumer
   (*b*) consumer      (*e*) producer
   (*c*) producer      (*f*) consumer

**8** Nutrition is the sum of the processes whereby an organism obtains energy and materials in order to maintain the processes of life.
Energy is the capacity to do work.
Trophic level is a feeding level within an ecosystem, for instance producers, primary consumers, secondary consumers.
Food web is the interrelationships between the food chains in an ecosystem.
An omnivore is an organism which feeds on both plant and animal matter.

**9** Inorganic nutrients – minerals, water
Organic nutrients – protein, carbohydrate, fat

## Self test 2

**1** Ingen-Housz

**2** (*a*) Van Helmont showed that vegetable matter was formed from water.
(*b*) Priestley showed that vegetation could restore injured air.

**3** (*a*) water + carbon dioxide $\xrightarrow[\substack{\text{green} \\ \text{vegetation}}]{\text{light}}$ vegetable matter + oxygen

(*b*) water + carbon dioxide $\xrightarrow[\text{chlorophyll}]{\text{light energy}}$ glucose + oxygen

$$6H_2O + 6CO_2 \longrightarrow C_6H_{12}O_6 + 6O_2$$

## Self test 3

**1** (*d*)    **2** (*e*)    **3** (*d*)    **4** (*a*)

**5** See figure 121.

### 121   Action spectrum of photosynthesis

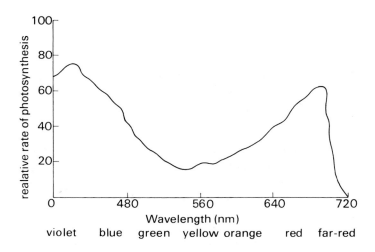

An action spectrum shows graphically the relationship between the wavelength of light and the rate of a light-dependent process such as photosynthesis. The action spectrum for photosynthesis shows that its maximum rates are in red and blue light.

**6** (*a*) An absorption spectrum shows the extent to which a pigment absorbs different wavelengths of light.
(*b*) The absorption spectrum for chlorophyll *a* shows a maximum absorption in blue and red light. Accessory pigments are able to absorb light in the middle region of the spectrum where chlorophyll's absorption is inefficient.

**7** Shade plants contain a high concentration of chlorophyll

*b* which has absorption peaks nearer the centre of the spectrum than chlorophyll *a*. This enables them to use those wavelengths of light which reach them. Much of the blue and red light has already been filtered out by the leaves of sun plants above them.

**8** Blackman found that at low light intensities photosynthesis has characteristics of a photoreaction, while at high light intensities it has characteristics of a chemical reaction. He therefore suggested that photosynthesis comprises two stages; a light-dependent stage and a light-independent chemical stage.

## Self test 4

**1** (*a*) $ADP + P_i$              (*d*) ATP
(*b*) $ADP, P_i + NADP$       (*e*) $ATP + NADPH_2$
(*c*) non-cyclic photo-        (*f*) $O_2$
phosphorylation

**2** (*a*) An electron which has been raised to a higher energy state. This happens to chlorophyll electrons during photosynthesis when light is absorbed by the molecule.
(*b*) The passing of electrons from one molecule to another. Such molecules are known as electron carriers. This occurs in photosynthesis when the excited electron leaves the chlorophyll molecule and passes along a series of electron carriers.
(*c*) The synthesis of the energy carrier molecule ATP from ADP and $P_i$, using energy from the high-energy electrons created during absorption of light by chlorophyll. This energy is released and coupled to ATP synthesis as the electron passes along a pathway of electron carriers.

**3** $H_2O$

**4** (*a*) $H_2O \rightleftharpoons H^+ + OH^-$

$2OH^- \rightleftharpoons H_2O + \frac{1}{2}O_2 + 2e$

(*b*) $H^+ + 2e$ – used in the formation of $NADPH_2$
$O_2$ evolved

**5** (*a*) Hill isolated chloroplasts and showed that when illuminated they produced oxygen and a reducing agent. This became known as the Hill reaction.
(*b*) Arnon developed a technique for isolating chloroplasts which he then showed were able to convert $CO_2$ to carbohydrate in the presence of light.

# Self test 5

**1** Both molecules are involved in the light-independent phase. PGA normally reacts with products from the light-dependent reaction, since absorption of light causes it to rise in concentration.

RDP is formed as a result of reactions which depend on products from the light-dependent reaction, since its concentration falls in the absence of light.

**2** See figure 122.

**122   Calvin cycle**

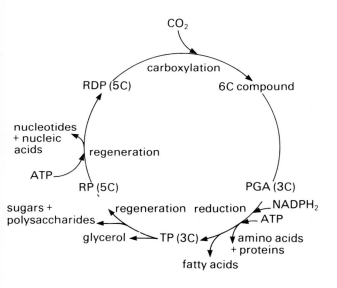

**3** The intermediates of photosynthesis were radioactively labelled by exposing algae to $^{14}CO_2$. The chemicals in the algae were then extracted and separated and identified by means of chromatography. Those chemicals which were intermediates in photosynthesis, and which were therefore labelled, were identified by the technique of autoradiography. When the chemicals were extracted at increasing time intervals after exposure of the algae to radioactive $CO_2$, the sequence of reactions of the Calvin cycle were worked out.

**4** A plant which can fix $CO_2$ into a 4-carbon acid as well as PGA.

# Self test 6

**1** (*a*)
$$6CO_2 + 12H_2S \xrightarrow[\text{bacteriochlorophyll}]{\text{light}} C_6H_{12}O_6 + 12S + 6H_2O$$

(*b*) The source is hydrogen sulphide. Bacteriochlorophyll is the pigment.

**2** Chemosynthetic bacteria obtain their energy by the oxidation of inorganic molecules such as ammonium compounds and ferrous salts.

**3** (*a*) *Nitrosomonas*
(*b*) $(NH_4)_2CO_3 + 3O_2 \rightarrow 2HNO_2 + CO_2 + 3H_2O + \text{energy}$

**4** Chemosynthetic bacteria are important in the recycling of mineral salts. These salts are required directly as nutrients by plants.

# Self test 7

**1** (*a*) A Epidermal cells, guard cells
B Palisade mesophyll cells
C Spongy mesophyll cells
D Xylem
E Phloem
(*b*) F would be found in D
G would be found in B
H would be found in A
(*c*) (i) is associated mainly with B (also C)
(ii) is associated mainly with E
(iii) is associated mainly with C
(iv) is associated mainly with A

**2** A (iv) and (v)
B (i), (ii), (iii)
C (vi)

## Self test 8

**1** An essential element is an element required by plants and without which they are unable to grow and develop. A trace element is an essential element required in only very small quantities.

**2** Water culture

**3** Nitrogen, phosphorus, sulphur, potassium, calcium, magnesium, iron, boron, manganese, copper, zinc, molybdenum, chlorine

**4** One from each of the following.
(a) nitrogen, sulphur
(b) nitrogen, magnesium
(c) calcium
(d) phosphorus
(e) nitrogen, sulphur, potassium, magnesium, manganese, copper, zinc, molybdenum, chlorine
(f) phosphorus, magnesium, manganese

## Self test 9

**1** Any three from each of the following.
(a) Goose barnacle - Arthropoda
*Amoeba* – Protozoa
Mussel - Mollusca
Cockle - Mollusca
Tellinid - Mollusca
Peacock worm – Annelida
*Daphnia* – Arthropoda
Flamingo - Chordata
Blue whale – Chordata
(b) Peacock worm – Annelida
Flamingo - Chordata
Blue whale – Chordata
(c) Frog - Chordata
Jellyfish - Coelenterata
*Arenicola* – Annelida
Earthworm - Annelida
Snail - Mollusca
Sea urchin – Echinodermata
Crab - Arthropoda
Sea anemone – Coelenterata
Large-mouth bass - Chordata
Cuttlefish - Mollusca
Octopus - Mollusca
Starfish - Echinodermata

(d) Elephant hawk moth - Arthropoda
Hummingbird – Chordata
Aphid - Arthropoda
Mosquito - Arthropoda
Young mammal - Chordata
Spider – Arthropoda
Tapeworm - Platyhelminthes

**2** (a) (i), (vi), (viii), (ix), (xv)
(b) (iii), (v), (vii), (x), (xiii)
(c) (ii), (iii), (xi), (xiv)
(d) (iii), (iv), (vii), (x), (xii)

**3** (a) Three from bacteria, algae, protozoa, rotifers, nematodes.
(b) It is sensitive to chemicals from its food.
(c) Pseudopodia
(d) In food vacuoles

**4** Any four from: –

| Starfish | Spider |
|---|---|
| (i) Pull open bivalve with numerous tube feet | (i) Pierce prey with hollow jaws (chelicerae) |
| (ii) Force everted stomach into bivalve and over soft tissues | (ii) Digestive juices are pumped into body of prey |
| (iii) Absorption of digested products occurs from stomach **in** prey | (iii) Exoskeleton of prey sucked empty – absorption **within** gut of spider |
| (iv) No specialised mouthparts for feeding | (iv) Specialised mouth-parts |
| (v) Slow pursuit or search for prey | (v) Builds traps for prey, or rapid movements to catch it |

**5** Labrum, mandibles, maxillae, palps, sensory, labium

**6** See figure 123.

**7** (a) Herbivore (actually a rabbit)
Any two from:
(i) Presence of diastema.
(ii) Cheek teeth to provide a relatively flat grinding surface.
(iii) Well-developed incisors (for gnawing)
(b) i. $\frac{2}{1}$    c. $\frac{0}{0}$    pm. +    m. $\frac{6}{5}$
(c) *Mandible*

**8** (a) Canines lie between incisor and premolar in both upper and lower jaw. Carnassials are formed by pm.$^3$ and m.$_1$.

**123 Section through mammalian tooth**

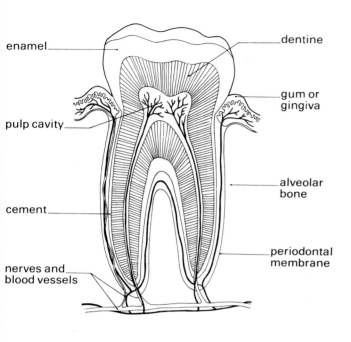

enamel

dentine

pulp cavity

gum or gingiva

cement

alveolar bone

nerves and blood vessels

periodontal membrane

(b) Canines are long, single pointed teeth (dagger-shaped). Carnassial teeth have the crown elongated into ridges, with two points.
(c) Canines have stabbing action. Carnassials act like blades of a pair of scissors to shear flesh and bone.

(b) Involved in the process of obtaining energy from food (respiration)

5 Any two from the following.
(a) Calcium, phosphorus, magnesium
(b) Iron, phosphorus, sulphur, potassium
(c) Sodium, potassium, chlorine
(d) Iron, phosphorus, magnesium

6 (a) As a structural component of cells and hence the body.
(b) As a medium for transport within the body
(c) As a medium for biochemical reactions

7 (c)

8 Four from each of the following.
(a) Protein, iron, vitamins A and D, riboflavin, nicotinic acid
(b) Protein, carbohydrate, calcium, iron, thiamine, nicotinic acid

9 (a) Vitamins $B_1$, $B_2$ and C, folic acid, minerals
(b) Oranges (Vitamin C and folic acid)
Wholemeal bread (Vitamin $B_1$ and folic acid)
Milk (Vitamin $B_2$)
Green, leafy vegetables (minerals and folic acid)

10 (a) Benedict's solution; sugars
(b) Structural component (skin, hair bone), movement (muscle), maintenance of metabolism (enzymes, hormones etc.); pink red
(c) Cloudy emulsion; lipid
(d) Energy supply and store, structural components (such as cellulose); iodine

# Self test 10

1 Lipids and minerals

2 (a) Kilojoule
(b) Basal metabolic rate
(c) (i) Child of 6 (ii) Breast-feeding woman (iii) Coal-miner

3 (a) (iii) (v) (vii)
(b) (i) (iv) (v) (vii)
(c) (i) (ii) (vi)

4 (a) (i) Vitamin C (ii) Vitamin D (iii) Vitamin $B_1$
(iv) Vitamin A

# Self test 11

1 (c)    2 (d)    3 (e)    4 (b)    5 (d)    6 (a)    7 (c)    8 (b)

9 (c) 10 (c) 11 (a) (NB pancreatic juice requires mixing with intestinal juice for trypsinogen to be activated.)

12 (d)        13 False        14 True

15 True        16 False        17 False

18 1 longitudinal and circular muscles
2 lumen
3 mucous epithelium

4 submucosa
5 serous coat

19 Mechanical digestion is the chewing action of teeth, churning action of stomach, etc.
Chemical digestion is the hydrolysis of complex food molecules by digestive enzymes.

20 Removal of waste food from body

21 (b) and (c)

22 (a), (c) and (d)

23 (b) and (d)

24 (a), (b) and (d)

25 (b), (c) and (d)

## Self test 12

1 Respiration is the process whereby organic molecules are broken down to release energy. This energy is required for the synthesis of new materials, the transport of substances within the body, movement, nerve transmission, etc.

2 (a) (iii)
(b) (ii)
(c) (iv)
(d) (i)

3 Although glucose contains much energy, the molecule itself is relatively unreactive. An external source of activation energy, in the form of ATP, is required before it can be made to release its energy.

4 (a) Pyruvic acid, cytochrome $b$, cytochrome $a$
(b) NAD, cytochrome $c$, oxygen
(c) NAD, cyctochrome $c$, oxygen
(d) pyruvic acid, cytochrome $b$, cytochrome $a$

5 (a) All three
(b) Cytochrome $a$
(c) FP (flavoprotein)

6 High-energy reduced cofactors, produced during respiration, pass electrons along a series of electron carriers. During this process, energy is released and coupled to the synthesis of ATP. Oxygen acts as the final electron acceptor in the series.

7 (a) Cytoplasm
(b) Matrix of mitochondria
(c) and (d) Inner membrane of mitochondria possibly associated with stalked granules
(e) Cytoplasm

8 See figure 98.

9 (a) 0.7 (57/80)
(b) A carbohydrate

10 (a) This is because the oxygen required for aerobic respiration cannot be transported to the cells fast enough.
(b) Lactic acid
(c) The extra oxygen is required by the muscle cells to oxidise the lactic acid accumulated during exercise.
(d) It can be used to phosphorylate ADP to produce ATP.

11 See figure 124.

## Self test 13

1 Compare your answer with figure 101.

2 The energy content of progressively higher trophic levels becomes progressively smaller. Thus, above trophic level 5, there is insufficient energy available to support organisms in any further trophic level.

3 A pyramid of energy is a diagrammatic representation of the energy content of each trophic level in an ecosystem (see figure 125).

The horizontal length of each block in the diagram is proportional to the energy content of each trophic level measured in joules per square metre per year.

4 (a) It makes no allowance for variation in size of organisms.
(b) It makes no allowance for variation in life-span or reproductive rate of organisms.
(c) It is a very time-consuming exercise.

5 Energy flows through ecosystems and cannot be recycled. Nutrients cycle through ecosystems and can be re-used.

6 Compare your answer with figure 105.

| Anaerobic | Aerobic |
|---|---|
| 1 Glucose, other carbo-hydrates, lipids and proteins | Glucose, other carbo-hydrates, lipids and proteins |
| 2 Partial | Complete |
| 3 Ethanol + $CO_2$ or lactic acid | $CO_2$ + $H_2O$ |
| 4 Cytoplasm | Cytoplasm + mitochondria |
| 5 Low | High |
| 6 Two | Thirty-eight |
| 7 (a) Yes | Yes |
| (b) Yes | No |
| (c) No | Yes |
| (d) No | Yes |
| 8 No | Yes |
| 9 $C_6H_{12}O_6 + 6O_2 \rightarrow$ $6CO_2 + 6H_2O$ | $C_6H_{12}O_6 \longrightarrow$ $2CH_3.CH_2OH + 2CO_2$ or $C_6H_{12}O_6 \longrightarrow$ $2CH_3.CH.OH.COOH$ |

125 Pyramid of energy

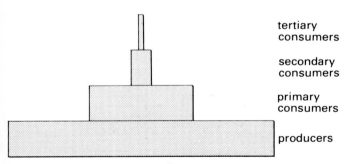

tertiary consumers

secondary consumers

primary consumers

producers

# Section 8  Answers to self-assessment questions

**1** (*a*) Energy (*b*) Nutrients (*c*) Nutrition

**2** (*a*) Energy supply (*b*) Energy supply, organic nutrient
(*c*) Inorganic nutrient (*d*) Inorganic nutrient (*e*) Inorganic
nutrient (*f*) Energy supply, organic nutrient

**3** (*b*), (*c*), (*d*)

**4** See figure 126.

**5** (*a*) Dead wood → insects → shrew → owl
Oak leaves → purple hairstreak → chiffchaff →
sparrowhawk
Hazel → insects → shrew → owl, etc.

**126   Food web in an oak wood**

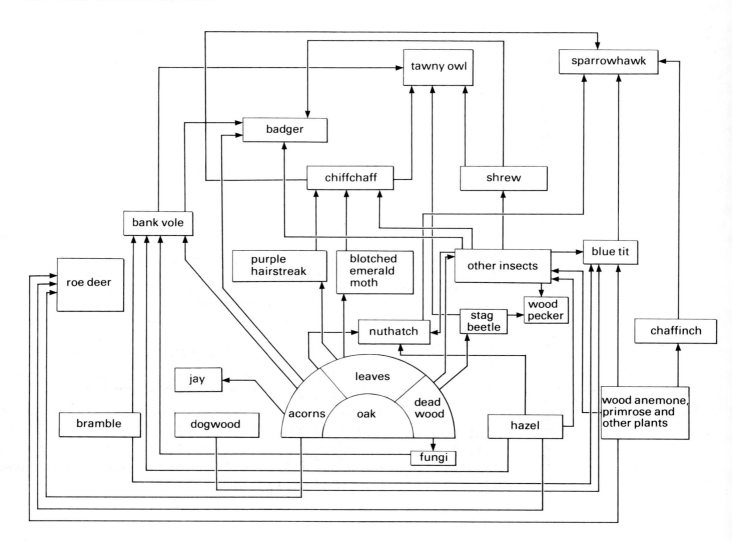

(b) Acorns → vole → owl
Dead wood → insects → badger
Dead wood → stag beetle → woodpecker, etc.
(c) Three of the following.
Sparrowhawk, tawny owl, badger, roe deer, woodpecker, jay
(d) Tawny owl, sparrowhawk, badger

**6** Stag beetle – detritivore
Chiffchaff – carnivore
Fungi – saprophyte

**7 Producers**
Oak, bramble, dogwood, hazel, wood anemone, primrose and other plants

**Primary consumers**
Purple hairstreak, blotched emerald moth, stag beetle and other insects, jay, nuthatch, bank vole, badger, roe deer, chaffinch, blue tit

**Secondary consumers**
Woodpecker, badger, chiffchaff, shrew, blue tit, sparrowhawk, tawny owl, bank vole, nuthatch

**Tertiary consumers**
Badger, sparrowhawk, tawny owl

**Decomposers**
Fungi (crumble cap and rooting shank), stag beetle and other insects

**8** A trophic level is a feeding level, such as producers, primary consumers, etc.

**9** Van Helmont did not realise the role of carbon dioxide from the atmosphere and ignored the two ounces difference in soil mass which may have been due to mineral uptake. At the time, the composition of the atmosphere was not understood. The small loss in soil mass could have been overlooked as experimental error.

**10** (a) Priestley's investigation was qualitative because he did not measure the factors he was investigating. He recorded observations such as 'candles burn very well'.
(b) Priestley's concern for controls was evident in his procedure as he would carry out a trial in which injured air contained no plant. He could then compare this with the parallel investigation in which the only difference was the presence of a plant. His concern for repeatability of his results was evident in that he repeated his investigations 'not less than eight or ten times'.
(c) Plants are able to 'restore' air which has been 'injured' by burning candles in it or by respiration of animals.

**11** Injured air $\xrightarrow[\text{green plants}]{\text{light}}$ restored air

**12** (a) (i) He states 'I attack the problems that can be decided by experiment and I abandon those that can give rise only to conjectures . . . . . '.
(ii) He compares two plants which are 'taken at the same degree of nutrients, from the same soil' and measures them under identical conditions.
(iii) He repeated the investigations many times.
(b) (i) Dry weight is the weight of organic material remaining after all water has been driven away or excluded.
(ii) Fixation is the inclusion of an inorganic molecule into the organic molecules of an organism.

**13** Photosynthesis is a process whereby green plants convert water and carbon dioxide into vegetable matter and oxygen in the presence of light and minerals.

**14** (a) 0.1% carbon dioxide is the optimum concentration because it produces the highest rate of photosynthesis without leaf damage.
(b) Because beyond 0.5% carbon dioxide there is no increase in the rate of photosynthesis and the plant is damaged.

**15** (a) Increasing the light intensity led to an increase in the rate of photosynthesis. The rate at which it increases slows down above 12 kilolux until it reaches a plateau.
(b) Beyond 20 kilolux, the rate of photosynthesis does not change with increasing light intensity.

**16** $CO_2$ levels available in the atmosphere are already being used to the maximum, that is it is the level of atmospheric $CO_2$ that is limiting the increase in rate of photosynthesis.

**17** (a) A is a shade plant, B is a sun plant.
(b) A would be part of the undergrowth (herb layer). B would be a tree.
(c) Compensation point for A is 2.
Compensation point for B is 3.

**18** (*a*) The bacteria move to regions of higher oxygen concentration. If it is assumed that oxygen production is an indication of photosynthesis, then the movement of bacteria to a region is an indication of the rate of photosynthesis of that region.
(*b*) When the whole area is illuminated, the bacteria group around those regions containing the spiral chloroplast (centre photograph). This indicates that the chloroplasts are producing the oxygen which attracts the bacteria, and thus chloroplasts are involved in photosynthetic activity.
(*c*) No. In the investigation in which red and green light were both used, the red was focused on the chloroplast and the green on an area lacking chloroplast. No comparison can be made between red and green light's effectiveness in photosynthesis on this basis. This investigation showed that it was not green light itself which was attracting the bacteria.

**19** The movement of the bacteria is an indication of the rate of photosynthesis. The bacteria are more densely distributed in regions subjected to certain wavelengths of light. This indicates that the rate of photosynthesis varies with the wavelength of light available.

**20** The extent to which light of different wavelengths is absorbed by chloroplast extract follows a similar pattern to the rate of photosynthesis at these different wavelengths. This indicates that the chlorophyll pigments could be involved in absorbing light energy for photosynthesis.

**21** Two from carotene, phaeophytin, xanthophyll and chlorophyll *b*.

**22** (*a*) In curves A and B the reaction can be seen to go faster. Thus, possibility (*a*) is contradicted.
(*b*) Curve B shows the reaction going faster at the same temperature. Thus, possibility (*b*) is contradicted.
(*c*) Curves A and B show a faster rate at 20 kilolux. Thus, possibility (*c*) is contradicted.
(*d*) Comparison of curves C and D shows that this possibility is the only one supported: an increase in temperature without any increase in carbon dioxide concentration has no significant effect on the rate.

**23** (*a*) Rate of photosynthesis was reduced
(*b*) (i) Rate was increased (ii) Rate was unaffected

**24** Oxygen and a reducing agent

**25** That the whole process of photosynthesis, including the conversion of carbon dioxide to carbohydrate, could be carried out in isolated chloroplasts.

**26** (*a*) It could be (i) re-emitted, (ii) converted to heat, (iii) used to cause an electron from chlorophyll to be excited or raised to a higher energy level.
(*b*) (iii)

**27** The energy is released and coupled to the synthesis of ATP.

**28** Non-cyclic electron flow

**29** Two from water, oxidase enzyme, pigments, electron carrier molecule

**30** (*a*) (*c*)

**31** (*a*) (*b*) (*d*) (*e*)

**32** (*b*) (*c*)

**33** (*b*)

**34** ter

**35** (*a*) I
(*b*) I and II

**36** It raises the energy level of electrons from the chlorophyll molecule.

**37** The energy level falls because some of the energy is released and used to synthesise ATP from ADP and inorganic phosphate.

**38** (*a*) $CO_2 \rightarrow PGA \rightarrow TP$

(*b*) (i) TP $\begin{smallmatrix}\nearrow G \\ \searrow RP\end{smallmatrix} \rightarrow$ RDP

(ii) TP $\rightarrow$ RP $\rightarrow$ G $\rightarrow$ RDP

(iii) TP $\rightarrow$ RP $\begin{smallmatrix}\nearrow G \\ \searrow RDP\end{smallmatrix}$

Any two of the above alternatives are possible. You may have thought of others in addition.

**39** Rise

**40** RDP

**41** (i) PGA → TP
(ii) TP → RDP

**42** Three 5-C RDP molecules have been regenerated and three carbon atoms have been used to synthesise products such as sugars.

**43** (*a*) Carbon dioxide, PGA, pyruvic acid, fatty acid, fat
(*b*) Carbon dioxide, PGA, pyruvic acid, Krebs cycle, amino acid
(*c*) Carbon dioxide, PGA, TP, polysaccharide

**44** (i) PGA
(ii) TP
(iii) Carbohydrates such as glucose, sucrose and starch
(iv) RDP

**45** These bacteria are found in sulphur springs, and in organic mud at the bottom of lakes and ponds where there is a good supply of hydrogen sulphide.

**46** Photosynthetic bacteria (*b*) (*c*) (*e*)
Chemosynthetic bacteria (*a*) (*e*)

**47** The shade leaf is much thicker than the sun leaf, and in particular has a much wider palisade layer.

**48** Photosynthesis occurs mainly in the palisade layer. Thus, the shade plant has a larger volume of photosynthetic tissue which will enable it to make maximum use of what light it receives.

**49** The shade leaf will have the lower compensation point. This means that the photosynthetic rate will exceed the respiration rate, that is there will be a net synthesis of sugars, etc., at a lower light intensity than for the sun leaf. This is important in an environment where light intensity is always low.

**50** (*a*) Nitrogen and sulphur
(*b*) Nitrogen, magnesium, iron
(*c*) Calcium
(*d*) Phosphorus
(*e*) Nitrogen, sulphur, potassium, magnesium, iron, manganese, copper, zinc, molybdenum, chlorine
(*f*) Phosphorus, magnesium, manganese

**51** Enamel, ivory

**52** Dentine, periodontal membrane, pulp cavity

**53** Cement, periodontal membrane

**54** (*a*) Protects jaw bone and teeth roots.
(*b*) Carry oxygen and nutrients to living cells of the tooth and remove waste products of metabolism.

**55** Drilling in the dentine of young people, and some adults, where this region is still living can cause discomfort. However, should the drill enter the pulp cavity, pain is felt as this region contains nerve endings.

**56** (*a*) (ii) (vii)
(*b*) (iv) (vii)
(*c*) (i) (vi)
(*d*) (iii) (v)

**57** i.$\frac{2}{2}$     c.$\frac{1}{1}$     pm.$\frac{2}{2}$     m.$\frac{0}{0}$  = 20

**58** Age, sex, occupation/activity

**59** Both meals are of roughly equal energy value but the snack meal is, in some respects, of higher nutritional value than the hot meal. It provides the same amount of protein, four times as much vitamin A and five times as much calcium, but less of some other nutrients.

**60** The iron content of the snack meal is rather low as is the vitamin C content. The cooked meal is low in calcium, vitamins A and D.

**61 Snack meal**
Fresh green salad vegetable, such as watercress would provide iron. An orange would provide vitamin C.

**Cooked meal**
A glass of milk would provide calcium.
Carrots would provide vitamin A.
Custard made with eggs would provide vitamin D.

**62** Liver, eggs, cheese and meat, if vitamin $B_{12}$ is deficient.
Offal, raw green, leafy vegetables, pulses, bread, oranges

and bananas, if folic acid is deficient. Vegetable oils and cereals if vitamin E is lacking.

**63** Inclusion of more vitamin A in the diet in the form of fish-liver oils, liver, kidney, dairy products, eggs and vegetable products such as spinach and carrots.

**64** $B_1$ thiamin, $B_2$ riboflavin, nicotinic acid, vitamin C

**65** Vitamin D
1 Maintains levels of calcium and phosphorus in blood.
2 Enhances absorption of calcium from the intestine.
3 Regulates interchange of calcium between blood and bone.

**66** Mouth, oesophagus, stomach, duodenum, jejunum, ileum, caecum, colon, rectum, anus

**67** Appendix

**68** Duodenum

**69** (*a*) Stomach (*b*) Colon

**70** (*a*) Serosa and submucosa
(*b*) Mucosa and submucosa

**71** (*a*) Decrease (*b*) Increase

**72** (*a*) Bread, because there was no mention of it after 3.45pm. Presumably it was no longer detectable.
(*b*) Partly digested – broken into small shreds, soft and pulpy – nearly all particles of meat disappear, become chymified and changed into reddish-brown sediment suspended in fluid with small floating white particles (coagulae).
(*c*) Digestion continued when a portion of the stomach contents was placed in glass container (vial). Thus, only food and gastric juice were present and the living stomach activity was excluded.
(*d*) To maintain stomach contents at near-to-body temperature.
(*e*) Vegetables, that is potatoes and turnips. Enzymes are not able to penetrate cellulose walls to digest cell contents.
(*f*) Lightish brown, thin jelly with sediment and small white floating particles plus some undigested material (vegetables).

**73** (*a*) Stomach (*b*) Small intestine
(*c*) Mouth (*d*) Stomach

**74** (*a*) Stomach (*b*) Mouth and small intestine
(*c*) Stomach (*d*) Small intestine

**75** Pepsin : protein → polypeptide
Trypsin : polypeptides → dipeptides

**76** (*a*) Bile (*b*) Liver

**77** Villi

**78** Crypts of Lieberkühn

**79** Blood capillaries and lacteals (lymph capillaries)

**80** The plasma membrane is greatly folded, forming microvilli. This will increase the surface area of the cell boundary and allow more food to be absorbed.

**81** Mitochondria. They are the site of energy release. They provide the energy for active transport of digested foods from the lumen of the ileum.

**82** Plant food is very hard to break down to release the usable nutrients. Chewing the cud permits mechanical breakdown of material. This is especially important after it has been acted on by enzymes and then regurgitated.

**83** To digest cellulose in their food, since they do not have the required enzyme cellulase themselves.

**84** Mouth → oesophagus → rumen → reticulum → rumen → oesophagus → mouth → oesophagus → rumen → reticulum → omasum → abomasum → duodenum → jejunum → ileum → caecum → colon → rectum → anus

**85** (*a*) Rumen (and reticulum)
(*b*) Caecum and appendix

**86** (*a*) Stomach wall (*b*) Ileum wall

**87** Hepatic artery – oxygen
Hepatic portal vein – products of digestion

**88** (*a*) Citrulline, arginine, ornithine
(*b*) $CO_2$

**89** Fats, fatty acids, glycerol, glucose, amino acids

**90** It could be converted to fat or glycogen in the liver, or passed to the muscle where it is converted to glycogen.

**91** Glucose, fatty acids, glycerol

**92** They could be converted to glucose, converted to protein, or converted to urea and eliminated via the kidneys.

**93** Glycolysis

**94** See figure 127.

**127 Table for question 94**

| Stage | Raw materials | Products |
| --- | --- | --- |
| Glycolysis | Glucose + ADP | Pyruvic acid + ATP |
| Krebs cycle | Acetyl coA | Reduced cofactors + $CO_2$ |
| Electron transport | Reduced co-factors + $O_2$ | $H_2O$ + oxidised cofactors |
| Oxidative phosphorylation | ADP | ATP |

**95** (a) Lactic acid or ethanol + $CO_2$ and ATP
(b) $CO_2$, $H_2O$, ATP

**96** It has been split into two 3-carbon molecules.

**97** It completes the breakdown of the 2-carbon remains of the skeleton to $CO_2$. This is another example of decarboxylation.

**98** In animals, the skeleton is not broken down further, it is just rearranged to produce lactic acid. In plants, one carbon is removed as carbon dioxide and 2-carbon ethanol is produced.

**99** $b \rightarrow a$

**100** Strongest

**101** Reduced

**102** It is an oversimplification as it implies that the oxygen reacts directly with the glucose. It gives no indication of intermediary stages nor the way in which the energy is released in small manageable amounts.

**103**

$$RQ = \frac{102 \text{ vols. } CO_2}{145 \text{ vols. } O_2} = 0.7$$

**104** (a) Lactic acid, creatine
(b) $CO_2$, phosphocreatine

**105** (b)

**106** Protein synthesis, nerve impulse transmission, active transport of molecules across membranes, muscle action during movement, etc.

**107** (a) Sun (b) Producer level (c) Photosynthesis

**108** (a) Primary consumer (b) Secondary consumer (c) Tertiary consumer (d) Producer

**109** (b)

**110** (b)

**111** (a) As heat
(b) An increased energy input to the producers would lead to an increase in their biomass. In turn, there would be more energy available to the primary consumers and their biomass would also increase. In time, an increased energy input to the producers would lead to an increase in the biomass of all trophic levels. (This assumes that there are ample supplies of nutrients to the producers.)
(c) The effect of eliminating carnivores from the ecosystem would be to increase the biomass of the primary consumers. (If you shoot all the foxes, the number of rabbits increases.) The increased herbivore population would make greater demands on the plants, and thus the biomass of the producers would decrease.
(d) (iii) A directional flow
(e) The total energy available to each successive trophic level decreases. Thus, after about the fifth level, there is insufficient energy available to support any additional levels.

**112** See figure 128.

**113** See figure 129.

**114** See figure 130.

The pyramids of biomass for the two ecosystems would all be similar to the 'classical' pyramid.

**128   Pyramid of numbers for an oak wood ecosystem**

tertiary consumers

secondary consumers

primary consumers

producers

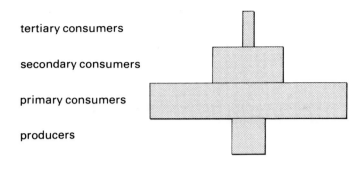

**129   Pyramid of numbers for a host/parasite food chain**

hyperparasites

parasites

host

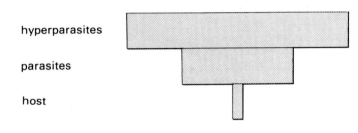

**130   Pyramid of biomass**

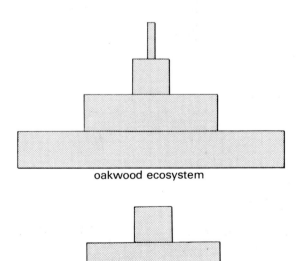

oakwood ecosystem

host/parasite ecosystem

**115** (*a*) By pyramids of energy
(*b*) Pyramids of number and biomass do not allow for difference in size and productivity of organisms.

**116** Energy flows through the ecosystem in a non-cyclic way, while nutrients are cycled.

## Pre-test: Carbohydrates, lipids and proteins

Answers        1 mark per question

**1**  carbon
**2**  hydrogen    } in any order
**3**  oxygen
**4**  glycosidic
**5**  one of the following: glucose, fructose, galactose, maltose
**6**  sucrose
**7**  aldehyde or keto (reducing) group
**8**  carbon
**9**  hydrogen    } in any order
**10** oxygen
**11** triglyceride
**12** fatty acid
**13** glycerol (propane-1,2,3-triol)
**14** carbon
**15** hydrogen
**16** oxygen    } in any order
**17** nitrogen
**18** sulphur
**19** amino acid
**20** peptide
**21** glucose    } in any order
**22** fructose
**23** maltose    } in any order
**24** sucrose
**25** starch
**26** glycogen
**27** cellulose
**28** fats
**29** oils    } in any order
**30** waxes
**31** glycolipids    } in any order
**32** phospholipids
**33** essential
**34** non-essential
**35** glucose
**36** starch
**37** glycogen    } in any order
**38** cellulose

**39** chitin
**40** fat around organs, e.g. kidneys
**41** subcutaneous fat in mammals
**42** cuticle in plants and insects
**43** skin ⎫
**44** hair ⎬ in any order
**45** bone ⎭
**46** muscle
**47** enzymes ⎫
**48** haemoglobin ⎬ in any order
**49** antibodies ⎪
**50** hormones ⎭

## Answers to programmed learning text post-test

| | |
|---|---|
| **1** (i) addition | 1 mark |
| (ii) removal | 1 mark |
| (iii) removal | 1 mark |
| **2** (i) removal of electrons | 1 mark |
| **3** (i) accepts | 1 mark |
| (ii) donates | 1 mark |
| **4** (i) reduced – $Cl_2$, CuO | 2 marks |
| (ii) oxidised – Na, C | 2 marks |
| **5** (i) A to F | 1 mark |
| (ii) Strongest reducing agent is A | 1 mark |
| Strongest oxidising agent is F | 1 mark |
| TOTAL | 13 marks |

# Index

rumen *68*
ruminants 68

saliva *62*
saprophyte *5*
scavenger *5*
secondary consumer 6
segmentation 58
shade plant 15, 21, 37, *37*
spectroscope 20, *20*
sphincter muscle 58
stalked granules 78, *79*
starch 12, *32, 56,* 61, *62*
stomata 34, *34, 36*
stroma *38*

succus entericus *see* intestinal juice
sucrase *62*
sucrose *32, 62*
sugar, non-reducing *56*
    reducing *56*
sulphur bacteria 33
sun plant 15, 21, 37, *37*

temperature effects, chemical reactions *22*
    photoreactions *22*
tertiary consumer 6
trace element 39, 52
triose phosphate *30*
trophic levels 6, 7, 82, *83*
trypsin *62*

urea 69, *69*

van Helmont, Jean Baptiste 8, *11*
vascular bundle 36, *36*
villus *66*
vitamin 49, *54,* 55

water 27, 49
water absorption 67
water culture technique 39
water oxidase enzyme 27

xylem 36, *36*